P9-DYZ-118

THE HOMESCHOOLING OPTION

PREVIOUS PUBLICATIONS BY LISA RIVERO

Gifted Education Comes Home

Creative Home Schooling: A Resource Guide for Smart Families

THE HOMESCHOOLING OPTION

How to Decide When It's Right for Your Family

Lisa Rivero

Fitchburg Public Library
5530 Lacy Road
Fitchburg, WI 53711

WITHDRAWN

palgrave
macmillan

THE HOMESCHOOLING OPTION
Copyright © Lisa Rivero, 2008.
All rights reserved. No part of this book may be used or reproduced in any manner whatsoever without written permission except in the case of brief quotations embodied in critical articles or reviews.

Material on pages 170–175 is adapted from "Secrets of Successful Homeschooling," by Lisa Rivero, *Understanding Our Gifted,* volume 15, issue 4. Reprinted with permission, Open Space Communications LLC, 800-494-6178, www.openspacecomm.com.

First published in 2008 by
PALGRAVE MACMILLAN™
175 Fifth Avenue, New York, N.Y. 10010 and
Houndmills, Basingstoke, Hampshire, England RG21 6XS.
Companies and representatives throughout the world.

PALGRAVE MACMILLAN is the global academic imprint of the Palgrave Macmillan division of St. Martin's Press, LLC and of Palgrave Macmillan Ltd. Macmillan® is a registered trademark in the United States, United Kingdom and other countries. Palgrave is a registered trademark in the European Union and other countries.

ISBN-13: 978-0-230-60070-6 hardcover
ISBN-10: 0-230-60070-0 hardcover

ISBN-13: 978-0-230-60068-3 paperback
ISBN-10: 0-230-60068-9 paperback

Library of Congress Cataloging-in-Publication Data
    Rivero, Lisa.
    The homeschooling option : how to decide when it's right for your family / by Lisa Rivero.
        p.   cm.
    Includes bibliographical references and index.
    ISBN 0–230–60068–9
    1. Home schooling—United States.   2. Education—Parent participation—United States.   I. Title.
LC40.R58   2008
371.04'2—dc22
2007025495

A catalogue record of the book is available from the British Library.

Design by Letra Libre

10   9   8   7   6   5   4   3
Printed in the United States of America.

To Al, for always being there, for the gelato, and for being "that kind of husband."

*To Albert, for every day and everything,
and for being who you've always been.*

*And to the Shakespeare Home Players:
You rock, fo shizzle! Allow no one to harsh your mellow.*

CONTENTS

THE HOMESCHOOLING OPTION

INTRODUCTION

"I could never homeschool my children."

If only I had a dime for every time I've heard this statement. I wrote *The Homeschooling Option* to show you that it is possible to homeschool, if you decide on that path for your family. It won't necessarily be easy all the time, but it can be rewarding. Your life will change. You will probably need to make some adjustments, especially at the beginning. Your neighbors or relatives might not understand, at least not completely. If you homeschool, you can expect transformations in your relationships with your children. You can also expect to explore and to question many of the beliefs you now have about education, family life, and society.

Scary? Maybe, but you've done this before. When you became a parent, you knew your life would change; you just didn't know by how much or in what ways. Becoming a homeschooling family is a similar adjustment. You go into it with certain expectations and for certain reasons, and soon find that it's not exactly as you imagined. You adjust. You learn as you go. You create a life of learning at home that is tailored to your children and family, and soon you begin to realize the rewards that home education can bring.

As homeschooling enters the public consciousness, education at home is becoming a real option for more families and easier than ever before. The wide diversity of homeschoolers, how they homeschool, and why they homeschool mean that they can no longer be dismissed as anomalies or a radical fringe group. A quick scan of recent newspapers and magazines shows that home education, with all its rewards, challenges, and questions, is a part of life in the twenty-first century.

An article in the *Journal of College Admission* notes that homeschoolers' ACT and SAT scores are higher than those of public school students, and home-educated college students perform as well as or better than traditionally educated students.[1] The *Boston Globe* reports that Ivy League schools and other top universities are welcoming homeschoolers with open arms.[2] For students who desire a college preparatory education, homeschooling can not only keep up with high schools, but, for some children with limited choices in public and private education—children in the inner cities, for example, or rural families—it may surpass what schools can offer. At the same time, home education advocates remind us that homeschooling is not just for gifted children, nor is its primary purpose to produce scholastic all-stars or spelling bee winners. Recent studies that compare homeschoolers' success with that of children in schools can leave parents wondering if home education is simply about getting an academic edge.

Realizing that homeschooled children are available during school hours, many museums, dance studios, ice rinks, libraries, bookstores, and other businesses offer classes for homeschoolers. The *Baltimore Sun* reports on homeschooled students in Maryland who use Irish dance and other community groups and classes to meet the state-mandated physical education requirements.[3] Where no such classes exist, parents are requesting them, and meeting with surprising success. Some homeschooling families have also found that there is such a thing as too much socialization, and they are cutting back on outside activities to spend more time at home.

An alternative to community- and business-sponsored classes and activities are parent-led homeschool co-ops, where children get together for learning and companionship. An article in the *New*

York Times reports that such co-ops are being started around the nation, as parents realize that they have the ability to create the social opportunities their children want and need.[4] At the Wisconsin Parents Association 2007 Home Education Conference, the session "Learn in the Company of Others While Homeschooling" showed parents how to work with other families to form book and writing groups, create science classes, and even take cross-country history excursions.

An article in the *Houston Chronicle* targets the homeschool consumer by offering home remodeling and decorating advice. Examples in the article reflect the diversity of homeschooled families and their approaches: some prefer a separate place for educational activities to keep home and school distinct; others make learning space and resources available in all living areas to integrate education and life.[5]

Publishers Weekly regularly reports on the market for homeschool books,[6] and bookstores and libraries stock scores of workbooks, guidebooks, curriculum guides, and other resources targeted specifically to the homeschool market. An article in *Kiplinger's Personal Finance* tallies the costs of homeschooling; it gives an overview of how extracurricular activities, distance learning, and quality curriculum materials affect homeschool families' budgets and offers some ideas for keeping costs low.[7]

This glut of information and resources is a mixed blessing for homeschoolers. While it is now possible to find almost any book, method, or activity to meet a child's need, parents also can find themselves lost in a maze of choices or lured by the promise that more money always leads to a higher-quality education. Some homeschoolers worry that the new homeschool consumerism strays from home education's roots of individual learning and family.

Contrary to stereotype, not all homeschoolers are in two-parent, one-income families with a stay-at-home mom. Some primary homeschooling parents are also employed, either in family businesses that give them more time at home or in jobs that allow for some flexibility of schedules. Some primary homeschooling parents are dads. Other parents have jobs that allow children to come to work with parents some of the time. A 2006 *Wall Street Journal* article takes a balanced look at the advantages and challenges facing two-income families who homeschool.[8]

Contrary to another stereotype, that blacks don't homeschool, African American families are drawn to homeschooling for the same reasons as other parents, including getting a rigorous education; many also emphasize the study of black history and a desire to prevent their children being unfairly labeled as needing special education. "There is an assumption that black boys are violent if they are too energetic," one parent says in a *San Francisco Chronicle* article. "My son is high-energy, and I didn't want him to end up on Ritalin or feel bad about himself."[9] A parent in a *New Pittsburgh Courier* article expresses similar concerns: "I want them to learn about the contributions made by Blacks to this country, and the world for that matter, more than a cursory look at slavery and the civil rights movement. . . . But most just don't want to put up with the subtle racism that exists in public schools. Parents I speak to say their sons are labeled slow or rambunctious and never given a chance to spread their wings."[10]

If these and other accounts of home education have piqued your interest, this book will help you to decide if homeschooling is the right option for you and your family. Not everyone or even a majority of families will decide to homeschool. As one parent I interviewed put it, "Imagine if someone decided to come up with the perfect diet for everyone on the planet, and, since it was the perfect diet, of course everyone had to follow it. I am very thankful that we have options so that everyone can choose for themselves." Homeschoolers don't want to force their choice on others. Even if you decide not to homeschool, knowing that the option is available and understanding how homeschool families learn and live—apart from the stereotypes and misconceptions—will change the way you view schools and education.

In decades past, non-homeschooling families probably didn't give a thought to home education, so the decision to use the school system was less of a conscious choice than a default position. Today, most parents have probably given at least a passing thought to homeschooling. Families who choose not to homeschool do so for many reasons. Some children enjoy and thrive in the classroom and simply do not want to homeschool. Some families are not willing or able to make the changes necessary for homeschooling to work well. Some parents do not have flexible jobs or schedules that allow an adult to be home (or the child to be at work) enough of the time. And other families find that family relationships are strained rather than strengthened when parents and children are together most of the time. All of these reasons for sending children to school are valid if they are well thought out and based on parents' understanding of schools, home education, and their own children's needs. Each family must choose for themselves, and no decision is irreversible.

Every homeschooler has a story. To give a broad picture of home education, I've drawn on my family's homeschooling and the experiences of several dozen other homeschool families throughout the years, both in our local homeschool support group and throughout the country. These homeschoolers' stories, while by no means statistically representative, give voice to what it can be like to learn at home. Two of the parents featured in this book are active in their state homeschool associations as coordinators for new homeschooling families. One parent has homeschooled five children, two have homeschooled three children, and another—who also works part-time—has a son at home and a daughter in public school. Two of the families have seen their older homeschooled children make a successful transition to college and beyond. The children and young adults range in age from fourteen to twenty-three, all old enough to have had time to reflect on their homeschooling experience. These families speak of a variety of approaches to and reasons for homeschooling. Some of the children homeschool by following their own interests, while others use a fairly conventional curriculum. Reasons the parents give for homeschooling include wanting to be able to watch their children grow, a desire for children to hold on to a love of·learning, accommodating special needs, wanting to spend more time together as a family and being able to travel, and having the freedom to practice their religion and observe holy days. Two of the six families whom I interviewed extensively—one Muslim and one Christian—cite faith-based reasons for their homeschooling; however, neither says that is the main reason.

Can you homeschool? Absolutely.

May you homeschool? Yes, in every state.

Do you want to homeschool? Only you know the answer to that question. This book will help you decide if education at home is a good choice for your family.

1

HOME, SCHOOL, AND EDUCATION

I am beginning to suspect all elaborate and special systems of education. They seem to me to be built upon the supposition that every child is a kind of idiot who must be taught to think. Whereas if the child is left to himself, he will think more and better, if less "showily." Let him come and go freely, let him touch real things and combine his impressions for himself.... Teaching fills the mind with artificial associations that must be got rid of before the child can develop independent ideas out of actual experiences.

— *Anne Sullivan, teacher of Helen Keller*

Are you curious about homeschooling? Do you wonder just how many homeschoolers there are, what they do all day, how they go about learning, and what happens to them after high school? Are you thinking about homeschooling but unsure of whether it is right for you, your child, or your family? Are you already home-schooling and looking for support for your decision, information to give to others, or more understanding of other homeschoolers?

Are you afraid of your children being socially out of step or academically unprepared?

If so, you are not alone, and this book will help to answer your questions. Homeschoolers are everywhere. Almost everyone knows someone who homeschools or who is thinking of homeschooling. While still a minority among families with school-age children, homeschoolers are definitely part of mainstream society. Every fall, newspapers and magazines run stories about homeschooling right alongside articles about going back to school, and these stories are not confined to the education pages nor do they always focus on the alternative nature of homeschooling. Museums, libraries, community groups, and bookstores offer homeschool memberships, discounts, classes, and other services tailored to children who learn at home. Most colleges accept homeschoolers; many have a separate admissions procedure for homeschoolers and actively recruit them.

In a relatively short period of time, homeschooling has gone from an idea on the margins to an accepted part of our educational system. Twenty-five years ago, many people didn't even know about the existence of homeschoolers, and home education was risky both socially and legally.[1] Today, it is legal in all fifty states, and while it might have suited only the unconventional family in the 1970s, homeschooling now appeals just as much to modern moms and dads who, until they had children, probably assumed that schools would be an unquestionable part of their lives. Homeschooling gives us all choices and control we thought we never had, and can change the way we think of education, family, and society. Homeschooling is here to stay.

WHAT IS HOMESCHOOLING?

When our son was younger and adults asked him where he went to school, he often said, "I don't go to school." I'm not sure if he noticed the quizzical—and sometimes alarmed—reactions he received. In his mind, not going school in no way implied he was not learning or being educated. Homeschoolers understand this. In *Family Matters: Why Homeschooling Makes Sense,* novelist David Guterson writes that a homeschooled child is simply "a young person who does not go to school." However, for people who have always equated school with education, one can't exist without the other. Eventually our son learned to say, "I am homeschooled," which, by the way, did not put an end to the quizzical reactions.

By homeschooling, we are exercising our right as parents to educate our child at home. All children in the United States are subject to compulsory attendance laws, and we fulfill the law by homeschooling. We take full responsibility for his education, and we take that responsibility seriously. Fortunately, it is a responsibility with rich rewards.

The right to homeschool in the United States is a basic constitutional freedom, and thus does not come from state or federal legislation. Larry and Susan Kaseman explain in a *Home Education Magazine* article that one common misunderstanding about homeschooling is that states grant families the right to homeschool by passing laws:

> We don't need laws to grant us the right to homeschool. At no point in history has homeschooling been illegal in any state. It is much better to keep reminding legislators and the general public of this fact than to ask the government to confirm it by passing a law, which then gives the government some authority over homeschools.[2]

Homeschooling is a parent's basic right that has been upheld in the courts. All school-age children are subject to compulsory attendance laws, not compulsory classroom attendance nor compulsory

education. This is a very important distinction. Compulsory attendance means that children must attend an educational program of
some sort, and this can include homeschooling. The government
cannot use compulsory attendance laws to say where a child will be
educated or legislate the kind of education a child must have. Even
schools are accountable for attendance only, not for proving that
they give all students an education.[3]

The U. S. Department of Education defines homeschooling as
follows:

> Homeschooled children may be taught by one or both parents,
> by tutors who come into the home, or through virtual school
> programs conducted over the Internet. Some parents prepare
> their own materials and design their own programs of study,
> while others use materials produced by companies specializing
> in homeschool resources. Accountability for homeschooling is
> coordinated with the state in which the family resides.[4]

Just as each state sets its own public school education requirements, every state has a legal procedure for homeschooling. Note
that this is different from the basic right to homeschool, which has
nothing to do with the law or the government. Prospective homeschoolers should become familiar with their state laws—discussed in
more detail in chapter nine—which can differ widely, and talk to
other local homeschoolers about how they meet state requirements.

OUR HOMESCHOOLING
STORY: THE BEGINNING

Our son's figure skating choreographer, a delightfully creative and
open-minded young woman, had already been giving him lessons
for several months when she learned that we homeschool. She said,

genuinely shocked, "You are the most normal homeschooling family I've met!"

I still chuckle when I remember that conversation. After eight years, learning at home is as normal to our family as waking up in the morning. Days, even weeks go by without my thinking consciously of ourselves as homeschoolers. Unless someone asks where our son goes to school or questions why he is out and about at the library or the grocery store on a school day, the topic rarely comes up. When people do learn that we homeschool, I'm sure many of them shake their heads and think that we seemed so, well, normal.

Some people know before they even have children that they want to homeschool, or they at least have the idea tucked away in the back of their minds. This was not our case. When our child started kindergarten, I knew no homeschooled children, and my only vague impression of homeschooling—when I thought of it at all—was that it was somehow undemocratic and for people who were "different" and wanted to separate themselves from society.

Also, I loved school! My first day of kindergarten, which happened also to be my sixth birthday, is one of those rare, treasured memories that I can call up at a moment's notice. With my new pencil case and Big Chief notebook in tow, I bounced through the door of the two-room rural schoolhouse where I was one of three children in my class. Like all students, I had good days and bad, but, overall, I looked forward to school and the learning I did there. I cried when school ended in the spring. I waited all summer to start a new grade in the fall. I knew from that first day that I wanted to be a teacher.

Since then, I've gone to public high school and private and public universities. Both my husband and I are college professors. I have nothing against schools, and this is not a book against public education or classroom teachers. Like Parker Palmer, author of *The Courage to Teach,* I agree that teachers today are "an easy target" and that "teacher-bashing has become a popular sport."[5] Being a classroom teacher and contributing in my own small way to the lives of young people is a privilege and a joy, and I have great admiration for the vocation and dedication of our finest elementary and high school teachers.

So why does our only child learn at home rather than in a classroom?

Like many families, we began homeschooling as an alternative to a conventional education that wasn't working. By the end of second grade, our son was changing from a joyful, curious, enthusiastic learner to a child who was unusually nervous, unhappily competitive, and generally miserable. The spark was gone. He was not failing in any measurable way—in fact, he was doing well academically—but he was definitely not thriving.

We decided to homeschool for third grade to buy us some time as we explored local private and public schools. Immediately after making our decision, our son insisted on starting homeschooling "right away" so that he could "start learning something." I hadn't realized how starved he'd been for a real education. Although we wouldn't be official homeschoolers until the fall, we thought of ourselves as beginning our journey that first summer.

Almost immediately, we saw our son recover his self-directed drive to learn and his joy and curiosity about everyday life. Without the worry about whether we'd find a school situation that would meet his needs, we could all relax for a while. I spent much of the summer thinking about the coming year and reading about home education. Most of the time I was excited and sure that we had made the right decision, relieved to see our son so happy and at ease with himself. There were also days when I questioned my sanity and wondered if perhaps we were expecting too much of our son or hadn't looked hard enough for other options, but we decided to wait a year before we reviewed our decision.

When everyone else in our neighborhood went back to school in September, we stayed home. Two other families whose children had gone to school with our son had also decided to homeschool that year, so we all got together once or twice a week. These get-togethers were crucial in my growing understanding of our new way of life and in our son's continuing to feel part of a learning community. The children often simply spent time together, especially when the weather was nice and they could be outside. On other days they did planned activities, such as building electrical circuits or doing nature studies. We always had time for food, fun, and conversation.

We three parents spent many hours at the kitchen counter, drinking coffee and discussing home education. Our individual approaches to homeschooling slowly evolved through the year to meet the needs of each family. One parent was very interested in child-directed learning and the role of internal motivation in education. She structured her home and her life to allow her children as much freedom as possible to follow their own interests and to tap into their internal motivation, all the while remaining in charge of and responsible for their education. Rather than use a published curriculum meant for homeschoolers, she preferred to create her own learning experiences for her children based on their questions and goals. Her approach to homeschooling reflected her personality, her parenting style, and her beliefs about education. She planned for a lot of wiggle room during the day, and she was open to changes of plans and creative bursts of energy.

The second parent gravitated toward a more structured homeschooling approach. She found that home education worked best within a detailed schedule and clear guidelines. She enjoyed her role as teacher, and she used literature units designed for homeschoolers and gave her daughters math assignments. Her family became involved with a large and active homeschooling community at their church. Again, her homeschooling choices grew from her parenting style, her personality, and her family's values.

My views were somewhere in the middle. I understood both approaches of my friends yet was hesitant to commit in full to either. Again, this said a lot about my outlook on life in general and our family members' individual temperaments. Our day-to-day homeschooling drew on several different educational theories. I learned much from the experiences and advice of others, and I've also learned to trust my intuition about our family's needs.

What I learned most from that first year and our conversations is that parenting style has a great effect on how families approach homeschooling. Many questions about homeschooling are really questions about parenting. For example, some parents are comfortable with their children setting their own times for going to bed and getting up, while others follow consistent schedules. Some parents expect their children to spend time each day on household chores or volunteer jobs, while others consider homework and school activities to be their children's work. For parents who homeschool, these kinds of choices inevitably become integrated with homeschooling decisions, and questions about how homeschoolers handle bedtimes or the balance between work, school, and play are as much a matter of how to parent as how to homeschool.

My experience is that the more challenging aspects of homeschooling are rarely about academics—what to learn, and when and how to learn it. While academic decisions are what scare many parents who are considering homeschooling, those questions are comparatively easy to answer. There is now a wealth of general education and homeschooling resources; if one approach or book doesn't work, another one probably will. With several national homeschooling magazines, hundreds of Web sites and Internet groups, local and state home education conferences, and a growing list of local support groups, homeschoolers can almost always find answers to such

questions as "How will I teach math?" or "How can my child prepare for the SAT?" The difficult challenges of homeschooling more
often deal with the intensive parenting that is inherent to home education, and with the initial discomfort of going against the societal
grain.

Homeschooling is intensive parenting because, quite simply,
parents who homeschool spend more time with their children. If
you are wondering what your homeschool days might be like, think
about the time you now spend together as a family—evenings,
weekends, summer vacations. More or less, that's how homeschooling will feel from a parent's perspective. For parents who already
have a good relationship with their children, the extra time offers a
chance to build on that relationship and to enjoy each other's company more. You will probably encounter some difficult issues that
you wouldn't otherwise, because they would have arisen in the classroom. Homeschooling parents are also more involved in their children's social interactions than parents whose children go to school.
For some parents, this means offering a higher level of emotional
and physical support to help children feel comfortable in groups.
For other parents, the challenge is learning what amount of distance
is appropriate for a specific child and finding ways to allow him or
her independence while still offering guidance for social skills.

Learning how to deal with perfectionism is one example of how
homeschooling helped me to be a better parent. From the time
when he learned to walk and to talk, our son has shown an innate
"urge to perfect."[6] Because of the nature of classroom learning, his
perfectionism quickly became a source of frustration at school,
where there wasn't always time to complete tasks and where his
motor skills often could not produce what was in his mind's eye.
When we started homeschooling, I needed to help him manage his
feelings and actions when he couldn't meet his own expectations. I
wanted our son to understand himself better, to accept who he is,

and to use his urge to perfect to his advantage rather than allowing it to be a source of anxiety. Perfectionism in itself is not a bad thing, and, if understood properly, it can be a source of pleasure as we work toward goals and envision what could be. This is a very different perspective from seeing perfectionism as a trait that can be eliminated or ignored.

With this in mind, rather than trying to change our son's nature, I focused on the power of parental example. I thought about how I might be modeling behaviors and attitudes that he was adopting as his own, even if I was not conscious of doing so. It's true that adults don't have meltdowns whenever they mess up. However, how often do we allow our children to see us dealing with and accepting the mistakes we do inevitably make? When children are in school, most of what we do is hidden from them, including our mistakes. Being at home gave me the chance to show our son with actions, not words, how to deal with perfectionist tendencies. I consciously began to let him see me make mistakes in my writing, cooking, and other daily activities, and to hear me talk myself through them ("Oh, well! This isn't a final draft, so I'll fix it later"). I played songs on the piano that were beyond my skill level, so that he could hear my mistakes and see how I dealt with them. In time, he became more tolerant of his own imperfections. Just as important, so did I.

Homeschooling doesn't magically fix shaky parenting skills or poor parent-child relationships. What it will do, however, is give you a valuable chance to improve your skills as a parent and to create the relationship with your children that you want. The slew of parenting books published every year shows how desperate we are to do right by our children. What none of these books can give us is time. The morning rush of preparation for work or school and evenings filled with homework and extracurricular activities leave little time to learn and practice new parenting skills. Many families today are in continual crisis mode, reacting to each problem rather than building a solid base of mutual understanding and respect so that fewer problems arise.

Being together all day, every day, forced us all to learn new and better ways to communicate and cooperate—valuable skills, but not always easy. Homeschooling did not absolve us from the trials and crises that all families face in one form or another; the difference was that we had the time and space to face those challenges, grow, and learn together. When a child goes to school, a parent who is going through a rough emotional time may be able to put on a stoic face until the school bus arrives, and then have the rest of the day to deal with life's ups and downs before everyone arrives home for the evening. Homeschooling parents and children see each other at their best and worst. When they have the time for open communication about any problems that arise, these problems become a source of growth, understanding, and respect rather than something to hide or run away from. As Fred Rogers was fond of saying, "Anything mentionable is manageable."[7]

Two additional challenges we have faced are being geographically far from homeschooled friends and juggling work and homeschooling. Even before we homeschooled, our son went to a small, private school that drew from a wide geographical area. His friends from that school lived as far as forty minutes away. As homeschoolers, we are in a similar situation in that our homeschooling friends do not live in our neighborhood. They live in adjoining towns and counties, and the parents must make an extra effort to get the children together and to arrange for rides and adult supervision. As some of the children get older and learn to drive, getting together has become easier, but it's not as easy as walking down the block or to the local playground.

Like a growing number of homeschoolers, ours is a two-income family. This has required some compromise on our part and some flexibility on the part of our employers. For the first couple of years of homeschooling, I took a break from college teaching and, instead, did limited freelance writing from home. When I did return to part-time teaching, my husband and I requested teaching schedules on alternate days, so that one of us was always home, especially when our son was younger. Our family is fortunate because both parents are able to work from home much of the time and we can set our own hours. When he got older, our son sometimes came with me to work and read in the library during my class. We've found it to be advantageous for him to have an awareness of our jobs and what we do for a living. His glimpse into the adult world of work gives him a better sense of where his eventual place there might be.

One seldom discussed aspect and benefit of homeschooling is that it teaches parents a lot about themselves. Not only do we learn alongside our children, but we also must come to terms with having chosen a way of life that—while more accepted than ever before—is still far from the norm. At the beginning, I was often self-conscious about being in public places with my son during school hours, and I wasn't sure how to answer questions about so-cialization or curriculum. In time, I learned to judge how much in-formation someone really wanted to know, and how much of our daily lives I felt comfortable sharing. Homeschooling helped me to be more secure in my own choices and to find a better balance be-tween considering different points of view and following our own path. Another homeschooling parent described how his commit-ment to home education paid off:

> Even on the worst bad days, my commitment to homeschooling never wavered. My wife, who has always been closely involved in our homeschooling on top of her demanding career, would some-times, out of concern for my mental stability, gently wonder whether our kids should be sent to school. But I never seriously considered it. If I could be certain about anything, I felt certain

even in those early years that homeschooling would continue to be best for my children. And now that we have almost ten years behind us, I can say, with more conviction than I would have ever thought possible, that my children have flourished as home-schoolers; it has been the best possible path for us.[8]

Even veteran homeschooling parents and older children say that dealing with how others perceive homeschooling is one of home education's biggest challenges. If we can work through this discomfort and use it to understand ourselves better, homeschooling can give us the self-confidence and courage to branch out in other areas of our lives as well, to try something new or to be willing to step away from the crowd.

HOMESCHOOLING LONG TERM

After we'd been homeschooling for a year, we reassessed our home-school experience, and we decided to continue. Elementary school turned to middle school, and, before we knew it, we faced high school. Once again, we wrestled with stereotypes and the idea that homeschooling may be fine for young children, but certainly teenagers need more structure if they are to get into college, broader social circles if they are to learn to get along in the world. Right?

I began to pay attention to the homeschooled teens we knew. They seemed unusually at ease in their own skins, confident about their abilities, and candid in their opinions. They had real conversations with their parents and with other adults. Their older siblings were going on to college after having homeschooled and reported no trouble with a transition to classroom education. I'm sure our son noticed these things about his friends as well. After visiting the local public high school and researching a couple of private high schools, he decided to homeschool for high school.

We've never looked back. Little did we know that first summer that with each passing year, homeschooling would feel less like a form of escape and more like the education option we'd been looking for all along. The idea of teaching one's children doesn't really capture how homeschooling works, because it implies a formal tutoring relationship rather than a family experience. In essence, homeschooling is no different from teaching our children to wash the dishes or write thank you notes or drive a car. Parents are already their children's teachers, and homeschooling parents continue in that role, just on a larger scale. Home education doesn't require a radically different set of skills or knowledge base. Knowing how to teach a room of twenty-five students is very different from working with one's own child, which we'd been doing all along. Also, as we will discuss in more detail in chapter six, homeschoolers teach themselves much of the time, and our job as parents is to facilitate, encourage, and support.

Over the years, our son's course of study has changed to meet his needs. We've learned that he can absorb an incredible amount of knowledge on his own in subjects such as history, literature, current events, the history and structure of language, and other humanities topics. In these areas, he is much better served by self-directed study through library books, Web sites, and conversation than by using a textbook or other formal curriculum. I help him with math and science, much of which we learn (or re-learn) together using a variety of resources. He had already begun to learn Latin in school, so his dad has continued a study of Latin with him at home, and our son's enjoyment of languages led him to take an ancient Greek class last year at a local university.

When he's not taking classes, working with his dad or me, or learning on his own, he participates in a homeschool teen literature and writing group, takes Tae Kwon Do classes, is part of a children's theater, and gets together with friends. Just as important, he has plenty of time for unscheduled, self-directed learning, and he continues to approach homeschooling each and every day with joy and enthusiasm. We place a high priority on making sure we aren't overscheduled, and we try to avoid feeling "too busy."

The reasons we now homeschool have little to do with what may be wrong with schools. They have everything to do with what works for our son and our family, what helps him to grow, what makes him a better student and a better young man, what challenges him, what fits his personality, abilities, and learning styles. I see our evolving perspective on homeschooling reflected in many other homeschooling families, some of whose stories you will read in the following pages. Homeschooling works for us—as it does for many other families—as a way to live as much as a way to learn.

2

HOMESCHOOLING PAST AND PRESENT

I think the biggest misconception is that homeschooling is the same as school, except that you are at home with your mom or dad teaching you. It might mean that to some, but, for many, "homeschooling" is really a misnomer because much of their learning is not at home at all. They may be out learning to fly an airplane or building a house. Some homeschoolers travel extensively and others spend time volunteering in the community. They meet and become friends with people of all ages. Home may be your base, but the "classroom" can be anywhere.

— Susan, homeschool regional coordinator

WORDS, WORDS, WORDS

State laws have various ways to describe and define homeschooling. Some states require homeschools to operate as private schools. Others use terms such as "home-based private education." Homeschoolers themselves may write homeschooling as one word or two, or prefer the phrase "home education" or "home learning."

The very word "homeschooling" can cause confusion, and is not always a good description of how homeschoolers live and learn.

John Holt, who was instrumental in the growth and direction of modern homeschooling, explained that homeschooling is not just another form of school:

> I have used the words "home schooling" to describe the process by which children grow and learn in the world without going, or going very much, to schools, because those words are familiar and quickly understood. But in one very important sense they are misleading. What is most important and valuable about the home as a base for children's growth into the world is not that it is a better school than the schools but that it isn't a school at all.[1]

The word "home" also doesn't always capture the essence of homeschooling. Many homeschoolers get much of their education away from home (in nature or groups or libraries, for example), and many others get their education in ways that bear little resemblance to school, as this former homeschooler remembers:

> *I'd try my best, for the most part, to be done with all the book stuff by noon. After noon was when we'd head out of the home. Homeschooling is a misnomer in that at least fifty percent of the time we weren't at home for our learning opportunities such as field trips, studying with other homeschoolers as a group for literature, science, politics, computers or just getting together for that all-important socialization factor.—Tim, age 23*

The socialization factor, which we will look at more closely in the next chapter, is a larger part of homeschoolers' days than is commonly thought. To reflect this fact, David Albert, homeschool parent and author, uses the phrase "community-based education," which, while lacking the simplicity of the moniker "homeschooling," describes how many families approach home education.[2] While a child in school has an occasional field trip outside the classroom, homeschooled children are in their communities nearly every day as they learn what is necessary to grow into the adult world. Homeschooled children learn from their parents, other family members, mentors in their neighborhoods, libraries, museums, the Internet, community classes, and each other, much as children have learned throughout the centuries.

HOW MANY HOMESCHOOLERS?

Statistics for homeschoolers' demographics vary widely. The National Center for Education Statistics reported that 1.1 million children (2.2 percent of all school-age children) were homeschooled in 2003, and the United States Census Bureau cites estimates of 2 million homeschoolers with a growth of 15 to 20 percent each year.[3] Kurt Bauman of the Census Bureau explains, however, that homeschooling statistics are hard to evaluate and interpret because of variations in how data is collected. On initial analysis, "we can't say a lot about the growth of the home schooling population," other than that growth did occur in the 1990s, and homeschooling has "established itself as an alternative to regular school for a small set of families."[4]

An obsession with homeschooling numbers is both misleading and misses the more important lessons of what homeschooling has to teach us. Predictions that the percentage of homeschoolers would continue to rise at a constant rate led to a backlash of fear among school administrators and professional educators concerned about the effect of fewer school students on the teaching profession and educational institutions. The fear extended beyond the classroom. Schools seem as much a part of our culture as baseball and apple pie, so much so that when parents choose not to send their children to school, their decision can be viewed as not only out of the mainstream but vaguely undemocratic and with more than a little suspicion. Without schools, grades, homework, and sports as topics of conversation, parents don't know what to talk about.

Fear of homeschooling overtaking the educational system is unfounded. The fact is that the much-publicized growth trend in homeschooling has, in fact, leveled off, and the percentage of school-age children who homeschool has not changed much in the past several years.[5] In Wisconsin, for example, where families file notices of intent to homeschool with the state Department of Public Instruction, 0.57 percent of the school-age population homeschooled in the 1989–1990 academic year, growing to 1.90 percent ten years later in 1999–2000. However, that percentage has since remained fairly steady, reaching a peak of 2.04 percent in 2002–2003, and, most recently, 1.97 percent in 2005–2006.[6]

Because of the high degree of parental dedication required and because relatively few people are willing to take a less-traveled path, homeschoolers will probably always be a self-selecting minority. Tim, a recent college graduate, looking back on his parents' decision to homeschool, understands that homeschooling can be challenging, but the rewards are many:

> *Homeschooling is not for everyone; it is a serious undertaking that any parent has the potential to do successfully. It takes dedication and creativity. It will not be easy, for you or your child, but it will be rewarding and fulfilling. I give thanks every day that my parents were willing to sacrifice as much as they did so that my siblings and I could be raised the way we were.*

BEFORE HOMESCHOOLING

Before our modern idea of homeschooling, children who learned at home were described as self-taught, home tutored, or simply not having attended school. Although homeschooling in the United

States as a modern movement began in the 1960s, the practice of getting one's education from family and community is as old as the idea of education itself and certainly much older than large-scale institutional schooling. As homeschool expert Linda Dobson writes, "If somehow we could help our caveman see into the future, he would regard government-sponsored schools as the variant, as would the majority of his descendants at least until the middle of the nineteenth century."[7]

Children throughout history have received their education at home in many ways, and many of the foundational ideas of American homeschooling can be found in much earlier times. In the seventeenth century, for example, John Locke urged parents to treat education as an honor, almost a kind of guilty pleasure:

> I have always had a fancy that learning might be made a play and recreation to children: and that they might be brought to desire to be taught, if it were proposed to them as a thing of honour, credit, delight, and recreation, or as a reward for doing something else; and if they were never chid or corrected for the neglect of it.

Locke went on to say that children can find learning to read and write so important and desirable that they would learn it from other children if parents were to keep it from them. He describes a young boy who was resisting being taught reading by his mother. Locke advised her not to treat reading as a duty or something to dread. Instead, he suggested that she allow the boy to overhear Locke and herself discussing "the privilege and advantage of heirs and elder brothers, to be scholars; that this made them fine gentlemen, and beloved by every body," but for younger brothers such as the boy who did not want to read, "they might be ignorant bumpkins and clowns, if they pleased." The plan was set in place, with the result that the boy begged to be taught, his internal motivation and desire for learning kindled.[8]

We needn't look far for examples of notable politicians, writers, and artists who were self-educated or taught at home. Abraham Lincoln, except for a few months in school, was entirely self-taught, mostly through reading. Author Agatha Christie received her education at home, and was encouraged by her mother to write. Artist Andrew Wyeth stopped going to school after third grade. None of these people were described at the time as homeschoolers, yet today that's what they would be called.

These are just a few of many examples in the long history of home education. Skeptics might argue that these gifted figures would have had success regardless of where or how they were educated, and that, for most people, compulsory schooling has resulted in a better education than if children were taught at home. However, according to the National Center for Policy Analysis, the literacy rate in the United States was higher in 1852, when the first compulsory school law was passed, than it is today.[9]

This fact may seem counterintuitive if we automatically equate schooling with education and literacy, or the growth of government schools with educational progress. However, colonists of the New World placed a high value on the education of the individual. As early as 1642, Massachusetts Bay Colony legislated that the heads of households were responsible for teaching their children and apprentices to read. In colonial America, education was a family and a community endeavor: "Students were taught by one and all— schoolmasters, parents, relatives, tutors, governesses, clergymen, physicians, artisans, shopkeepers, lawyers, and even indentured servants." In the decades after the American Revolution, when the expanding book publishing industry made printed literature more widely available than ever, the role of the family in teaching children, promoted as "fireside education," became even greater. During this time, schools did have a place in childhood education, but "only a small part."[10]

MODERN HOMESCHOOLING BEGINS

Compulsory schooling changed both where and how children learned, and not everyone agreed that the change was a success. Writers from both sides of the political aisle have questioned whether schools always provide the education our children deserve and our society needs. Ivan Illich, in his 1971 book *Deschooling Society*, argued that in government schools, the process of education takes precedence over the substance of learning, and the "pupil is thereby 'schooled' to confuse teaching with learning, grade advancement with education, a diploma with competence, and fluency with the ability to say something new."[11] This "institutionalization of values" (promoting group-think rather than the individual's search for truth) affects everyone, but especially the poor, who have relatively few educational options.

From a very different perspective, economist and Nobel laureate Milton Friedman wrote, "There is no respect in which inhabitants of a low-income neighborhood are so disadvantaged as in the kind of schooling they can get for their children." He wondered what changes a "truly free-market educational system" might bring and how we can bring American education up to date with modern needs: "We essentially teach children in the same way that we did 200 years ago: one teacher in front of a bunch of kids in a closed room."[12] We go to school mainly because we think we need to go to school and it's just the way things are, not because school necessarily serves the purposes of education.

While Illich called for a disestablishment of schools and Friedman for a restructuring of the current system, both asked the same fundamental questions and expressed what Mitchell L. Stevens calls the "profound uncertainty about public schools' ability to manage" education of the individual.[13] This uncertainty helped to set the stage for homeschooling's appeal.

John Holt and Unschooling

> Of two ways of looking at children now growing in fashion—seeing them as monsters of evil who must be beaten into submission, or as little two-legged walking computers whom we can program into geniuses—it is hard to know which is worse, and will do more harm. I write this book to oppose them both.[14]—John Holt

Perhaps more than any other writer or activist, John Holt defined homeschooling as it is now practiced in the United States. Holt was a teacher who wanted to reform schools, wanted to change "the way schools deal with children." His first books, *How Children Learn* and *How Children Fail,* focused on child development and what he saw as the child's natural impulse to learn, and how that impulse can be nurtured or thwarted. In *What Do I Do Monday?, Escape from Childhood, Freedom and Beyond,* and *The Underachieving School,* Holt argued that children should be given the time to learn from and correct their own mistakes, tested only when they request it, allowed to experience emotion in the learning process, and encouraged in their use of fantasy.

When change within schools didn't happen, Holt shifted his focus to helping parents take control of their children's education. In 1976 Holt wrote *Instead of Education: Ways to Help People Do Things Better,* and the next year he founded the magazine *Growing Without Schooling,* which was the first homeschooling magazine in the United States.

John Holt wrote in *How Children Learn,* "All I am saying in this book can be summed up in two words: Trust Children. Nothing could be more simple, or more difficult."[15] Trust in the child's ability to learn is the bedrock of Holt's views on homeschooling, and he introduced the term "unschooling" to describe this kind of natural learning that happens outside schools. Holt argued that all of us have an innate desire and need to learn, not just as children, but for our entire lives. Schools tend to interfere with this innate drive with artificial expectations, evaluations, and a template of what education should be for everyone at a certain age. Families who follow Holt's ideas don't worry much about learning specific subjects at specific ages, or covering all aspects of math or science. Instead, they trust that when a child discovers the need to learn, he or she will find a way to learn, or will ask for guidance from a parent or other adult.

For several years after Holt's coining of the term, "unschooling" was used to mean home education. Not until the 1980s was the term "homeschooling" used to describe all full-time education that took place outside of conventional classrooms. As homeschooling became more widely chosen and diverse, the word "unschooling" began to be used to refer specifically to the homeschooling approach of Holt and others who promoted child-directed learning. In this book, when parents or children use the word "unschooling," it means home education that is based on and guided by the child's interests and initiatives rather than a conventional curriculum. Unschooling is facilitated by parents and other adults, but it is neither assigned nor directed by them.

Some faith-based homeschool parents struggle with combining an unschooling educational philosophy and traditional, more highly structured approaches to Christian homeschooling. Teri Brown, author of *Christian Unschooling*, explains, "I felt estranged from my Christian homeschooling friends because of my educational philosophies and isolated from many of my unschooling peers because of my faith."[16] To address this need, in the 1990s, Mary Hood and Teri Brown expanded the idea of child-directed learning to include Christian approaches to home education. Becca, a homeschool parent of three boys, says, "Relaxed Christian Homeschooling, a term coined by Mary Hood, was instrumental in the early years of my homeschooling journey. One of Mary's ideas is that you are a family, not a school. That really resonated with me."

It is clear that unschoolers, like all homeschoolers, are a diverse group. This has prompted some homeschoolers to use the term "radical unschooling" to refer to education that is entirely child led and child directed, to distinguish their educational philosophy and methods from a broader, less ideological practice of unschooling. As in every human arena, some people thrive on boundaries and principles to follow more than others do, even if those boundaries have to do with having no boundaries.

The impact of John Holt on modern American homeschooling cannot be underestimated. Holt died in 1985, and his final book, *Learning All the Time,* was published posthumously. While the magazine *Growing Without Schooling* ceased publication in 2001, Holt Associates continues to work with homeschoolers through book publishing and speaking engagements.

The Moore Formula

Dorothy and Dr. Raymond Moore also were prominent homeschooling figures in the 1980s. Their 1976 book, *Better Late Than Early,* argued for delaying formal education until ages eight to twelve, or even longer in some cases. Then, when children reach the age when they are developmentally ready for formal education— when they reach, in the Moores' term, an "Integrated Maturity Level"—they are able to catch up to their schooled peers very quickly. One way to think of this approach is that it extends the natural learning between child and parent that takes place in the preschool years—learning to walk, talk, and take care of oneself, learning about the immediate world through curiosity and personal interaction—through early elementary school.

The Moores were politically active on behalf of homeschoolers, and they warned against the dangers of using a rigid and strict curriculum with young children. Their Moore Formula for homeschoolers focuses on study, work, and service: for children, daily study for a period appropriate for the child (from minutes to several hours), daily physical work equal to the time spent on study, and daily service in the home or community; for parents, being a good example and heeding the child's interests and needs.[17]

At a time when the American education debate included prominent voices advocating earlier and earlier formal education, the Moores were encouraging parents to have their children skip kindergarten and the first grades altogether in favor of individualized learning in the home. The Christian-focused Moore Foundation continues to educate families in the Moore Formula of education, with an emphasis on keeping costs and stress low for parents.

Here is one parent's experience of homeschooling in the "early days," which shows John Holt's influence as well as the Moores' idea of delay of formal instruction:

> *When my oldest child was three months old back in 1979, I ran across an interview with John Holt in* Mother Earth News *in which he discussed homeschooling. I had always done well in school but related immediately to what he said. He advocated letting children learn concepts thoroughly, without peer pressure or artificial stopping times. His confidence in the desire and ability of children to learn was striking. . . . Our early homeschool years involved lots and lots of reading aloud, hands-on activities, music, and play. Overt work time gradually seems to increase as the children grow in their abilities. I try to challenge the children at their specific levels. It becomes easy to tell when material is too easy for a child, and frustration for both parent and child sets in when material is too difficult. There always seems to be an almost palpable change in the kids around age twelve or thirteen, when they are suddenly ready for more concentrated work and more complex thoughts. We use a number of textbooks, but also use the library extensively.—Alice*

HOMESCHOOLING COMES OF AGE

Inspired by writers such as John Holt and the Moores, homeschoolers in the 1970s and 1980s were pioneers who worked to change both the laws and public opinion. Contrary to common thought, these early homeschoolers were ideologically mixed, and included progressive-education proponents on the left and conservative Christians on the right. What they had in common was a desire to make home education easier for those who wished to take it on, better understood by the general public, and supported by legislation in every state.

At that time, families who educated their children at home had little legal recourse to prove their right to do so. The culture may have wanted education that was more focused on the individual, but when homeschoolers made this idea a reality, their children were often considered truants, and schools took parents to court to get the children back in the classroom.

In 1983, the Home School Legal Defense Association was founded by Michael Farris and Mike Smith "to defend and advance the constitutional right of parents to direct the education of their children and to protect family freedoms." Other homeschooling advocacy groups, such as the Wisconsin Parents Association, also were founded at this time and worked for unrestrictive legislation to protect the freedoms of homeschooling. By 1993, every state had laws that recognized the right to homeschool.[18] Homeschooling's strength at this time came from its diversity and the persistence of its proponents. Without the efforts of this first wave of homeschoolers, we would not have the options available to parents today.

Once homeschooling was officially legal and families did not need to fear criminal repercussions, the stage was set for homeschooling to thrive and to secure its spot in American education.

The social change that John Holt worked for has, in many ways, come to pass.

STRENGTH IN DIVERSITY

Homeschoolers may not have the strength of huge numbers, but they do have the strength of diversity. This diversity surprises many people, even many homeschoolers themselves, who soon find that families homeschool for many more reasons and in many more ways than they had assumed. The homeschooling group our family belonged to for several years includes Christians, Muslims, Jews, atheists, African Americans, European Americans, upper-middle-class families, low-income families, and families from many different suburbs and outlying rural areas—far more diverse than our local public school population. The one thing that brought us together was our choosing to educate our children at home, a choice I always appreciate more deeply when I see the variety of people for whom homeschooling works.

The Home School Legal Defense Association explains, "without this diversity, home schooling would still be a marginal movement."[19] Writing about homeschooling in the United States from the perspective of the United Kingdom, Paula Rothermel has argued that the heterogeneous nature of homeschoolers gives them a new, more visible role, "taking them beyond the early 1980s' caricatures of them as hippies and religious fanatics, beyond the 1990s' petty differences and squabbles, to a position of empowerment, whereby their differences and diversity are becoming their strength."[20]

The fact that in any given crowd almost anyone could be a homeschooler makes homeschooling more acceptable to the general public. The more people who know homeschoolers personally, the more they see that homeschoolers are less different from them than they imagined, and the more homeschooling is accepted as a normal activity rather than a threat to or a rejection of families who choose to send their children to school. Mitchell L. Stevens describes modern American homeschoolers as "visible actors in public space."[21]

Kurt Bauman of the Census Bureau posits that the appeal of homeschooling goes beyond a concern with the individual and extends to parents' wanting to "reclaim" and "make schooling valuable," to regain control in the face of high-stakes testing and national standards.[22] When families entrust their children to schools, they are giving up control over not only six or more hours of the child's day, but what subjects the child will study, how those subjects will be taught, how well the child's approach to learning will be understood and accommodated, and whether and how much the child will be tested. For many families, especially if they see their children struggling or not thriving in school, this loss of control feels like an abdication of parental responsibility. They feel a duty to provide their children with the education they deserve.

As with so many other cultural movements, American homeschooling is having an effect on the rest of the world. Because the United States has already done much of the work to make homeschooling understood and accepted, this process in other countries will be faster and easier, since less effort needs to be spent on laying a foundation.[23] Most international home-education stories include some comparison with or nod to the popularity of homeschooling in the United States.

Home education in the United Kingdom has received considerable media attention in recent years. Reliable statistics on British homeschoolers are even harder to come by than those on American homeschoolers. In the United Kingdom, home education is legal. If children never attend a school, the parents do not need to notify the government of their intent to homeschool (although notification is required if children leave school for homeschooling). This leads to wild speculation about the numbers of homeschoolers, ranging from a low of 7,400 to a high of 150,000 children, although more reasonable estimates are between 25,000 and 50,000.[24] Other European countries can be divided into three groups in terms of their attitude toward homeschooling. The following countries, like the United Kingdom, have always permitted home education: Belgium, Denmark, Ireland, France, Italy, Luxembourg, Norway, Portugal, Sweden, and most of Switzerland. Austria now permits home education but did not do so in the past. Some countries do not technically permit home education, but seem to allow for individual cases to be approved: Spain, Greece, parts of Switzerland, the Netherlands, and Germany.[25]

Home-education laws in the rest of the world vary considerably. In some nations it is illegal (Brazil, for example). Like the United States, some countries delegate home-education legislation to states, territories, or provinces (Australia and Canada). Other countries, such as Japan, have vague laws and find themselves in the same situation that the United States was in twenty-five years ago, with a need for clearer legal definitions.

WHY HOMESCHOOLING WORKS

The reasons why homeschooling works are as varied as the families who homeschool. In chapter four, we will see examples of what a homeschooled education looks like. First, however, let's look at why homeschooling works for so many families and children.

Homeschooling Works Because It Allows Children to Learn at Their Own Pace

Homeschooling gives children the opportunity to soar in their areas of strength and take their time with more challenging subjects. Schools and teachers have little choice in how to manage large groups of children efficiently. To move a class in an orderly fashion from one grade to the next, they must treat everyone more or less the same. All third graders study the same science topics, read the same books, and learn the same math concepts. Anyone who is not yet ready will struggle, and a child who already knows the material will have to wait, patiently or not. Even with tracking and ability grouping, there is little chance of truly individualized learning in the sense of tailoring an education to a child's unique development.

In home education, the course of study can be designed around the child's needs, rather than trying to fit the child into a one-size-fits-all course of study. Does it really matter if a student learns cursive in fifth grade instead of third grade, for example? Or moves ahead in science even though his or her math skills are at grade level? Homeschooled children who naturally learn to read a bit later than average need not feel inadequate or behind, and those who learn quickly can cover more material without formally skipping grades. Home education accommodates a child's natural learning curve and eliminates artificial expectations.

Homeschooling Works Because It Provides a Safe Learning Environment

A vast majority of homeschoolers say that concern about the school environment—specifically drug use, safety, and peer pressure—is one of their reasons for homeschooling, and in the United Kingdom, bullying has been cited as the main reason that parents give for choosing home education.[26]

Are these parents paranoid or overprotective? Consider these findings on American schools in 2005, reported in a press release from the United States Bureau of Justice Statistics:

- Twenty-eight percent of students twelve to eighteen years old reported being bullied at school during the six months prior to the survey. Of those students who reported being bullied, 24 percent reported that they had sustained an injury as a result of the incident.

- Twenty-four percent of students reported that there were gangs at their schools, a 3 percent increase from 2003.

- Nearly all students twelve to eighteen years old encountered at least one security measure at school. The percentage of students who observed the use of security cameras at their schools increased from 39 percent in 2001 to 58 percent in 2005. At the same time, 90 percent of all students reported seeing school staff members or other adult supervisors in the hallway, and 68 percent of students reported the presence of security guards or assigned police officers at their school.[27]

When our son visited the local high school as an eighth grader, he entered through a metal detector. Many school children around the country do the same every school day. School violence and harassment have become so much an accepted part of school culture that this kind of surveillance is meant to be reassuring. While we can't and shouldn't try to shelter children from the reality of life and the world, many homeschool parents want to give their children a learning environment that is both physically and emotionally safe without the need for harsh reminders of violence.

Homeschooling Works Because It Strengthens and Nurtures Families

In a fast-paced culture that often pulls family members in different directions, homeschoolers have more time to live as a family rather than as roommates, chauffeurs, and passengers. Strong family relationships require, above all else, time—and time is the one thing over which families today feel they have little control.

Consider what might be a typical day for an American family with three children, ages ten, fourteen, and sixteen. All family members wake up early. The high school sophomore leaves the house at 6:45 A.M. to catch the school bus. One parent drives the middle child to school, while the other parent waits with the ten-year-old for his bus. Then both parents head to their jobs, and everyone is at school or work for most of the day. The high school student arrives home first, grabs a snack, then leaves for soccer practice, followed by drama rehearsal. The dad makes it home in time to meet the ten-year-old when he gets off the bus, then picks up the fourteen-year-old from school to take her to ballet class. In the lobby of the dance studio, the dad helps the ten-year-old with homework. The three of them stop for deli sandwiches on the way home. Meanwhile, the mom has arrived home after staying late for work. She eats, then picks up the high schooler from drama rehearsal. The dad drops off the fourteen-year-old, then takes the ten-year-old to ice hockey practice. They get home at ten o'clock. The high school student is in her room doing homework, which will take her until midnight to complete.

Does this sound exaggerated? It's not. Ask any parent who has active school-age children and you will hear a similar tale, if they can find the time for a chat and coffee. School sports and activities are often year-round rather than seasonal. Homework at competitive schools can take three or more hours per night. High school students are applying to six, twelve, or more colleges, requiring multiple tests and applications, interviews, and portfolios. Not just our schools and our children, but our entire culture is caffeinated, running at full speed so as not to miss anything or fall behind. It's as though we are on a merry-go-round, going faster and faster, and we are hanging on for dear life as life blurs around us.

Families on this merry-go-round often feel trapped, with no way to get off without sacrificing income, being unfair to one or more children by not offering each of them the same opportunities as their siblings, or feeling guilty for not helping their children live up to their potential. There is little time to catch one's breath, much less see things clearly enough to know if anything should be changed. Something is wrong with the pace of our daily lives when we need magazine articles and news reports to show us how to make time to be with our own children, how to eat together, how to schedule family time.

Homeschooling offers a way off the merry-go-round. We simply step off. In time, the world stops spinning and we can see life, education, and our children more clearly. We slow down and cut back. New and prospective homeschoolers are wise to read the works of John Holt and other early homeschool proponents who understood that homeschooling is not about giving an academic edge or riding a new merry-go-round. Homeschooling is about families being happier and less stressed, about getting to know each other and learning to live and learn together. Your children can still be well educated and successful; they'll just have more fun and be healthier doing so.

Homeschooling Works Because All Children's Needs Are Worthy of Special Attention

John Dewey argued nearly one hundred years ago that education starts with the child's unique needs.[28] Because each child has different needs, each homeschooler can become educated in an individualized, personal way. Some children's needs are harder to meet than others and require more work and knowledge and, sometimes, patience. However, each homeschooled child is in a position to have his or her unique needs considered special, with or without labels or diagnoses.

Some families decide to homeschool—either long term or temporarily—to address specific challenges such as autism, attention deficit disorder, or giftedness. Researchers reported in *School Psychology Review* that parents who homeschool children with attention deficit hyperactivity disorder (ADHD) "engaged their students at higher levels of academic responding than the public school instructors achieved with their students," although the parents had no formal teacher training.[29]

Parents of gifted children find that homeschooling removes barriers to deeper and broader learning, eliminates uncomfortable and unhealthy classroom competition from the learning environment, and allows sensitive children to get used to a busy, noisy world at their own pace. Homeschooling families with gifted children have created several popular online discussion groups, and some parents have formed support groups and learning co-ops specifically for such families.

Perhaps most important, each homeschooled child—regardless of labels or diagnoses—can be given an education tailored to his or her needs. While it is important for parents to understand how their children learn best and any difficulties they might have, there is no need to limit our understanding of children or their understanding of themselves with labels. This freedom to address each child as a unique learner is especially valuable when children within one family all learn differently. Treating each child's needs as special to that child is true equity in education.

DOING YOUR HOMEWORK: SELECTED READING

The following books can help all parents—regardless of whether they homeschool—to have a healthier, saner attitude toward the role of elementary, middle, and high school education in their child's lifelong learning:

Guerrilla Learning: How to Give Your Kids a Real Education With or Without School, by Grace Llewellyn and Amy Silver (John Wiley & Sons, 2001). This delightful book is a down-to-earth guide for parents whose children are in school. The goals of the book are not to show readers how to produce better grades or higher test scores, but rather to help parents view and approach education differently. It encourages parents to stop worrying so much about school performance, shows how children can become interested in learning and fully educated, and helps adults be more curious and better learners themselves. Grace Llewellyn is the author of the popular *Teenage Liberation Handbook,* and she draws on her many years of experience of working with homeschoolers.

What the Rest of Us Can Learn from Homeschooling, by Linda Dobson (Three Rivers Press, 2003). Another book from the homeschooling world written for school parents, this is a clear overview of what Dobson calls the principles of successful learning: curiosity, learning for the fun of it, wanting to succeed, practice, real-life learning, and intrinsic motivation. For parents who aren't sure if they will homeschool but who want to adopt a homeschooling mindset, this book offers a good way to stay centered and to sift through what matters in education and what doesn't.

Raising Lifelong Learners: A Parent's Guide, by Lucy Calkins and Lydia Bellino (Addison-Wesley, 1997). This is an insightful guide for parents who want to make learning an integral part of their children's lives. The authors stress that lifelong learners are nurtured at home first and foremost, and that parents have much more control than they think over their children's attitudes toward education. The book contains many practical examples and is useful for both homeschooled families and families with schooled children who want to nurture a more joyful attitude toward learning.

"Doing School": How We Are Creating a Generation of Stressed-Out, Materialistic, and Miseducated Students, by Denise Clark Pope (Yale University Press, 2001). This book follows the daily lives of five high school students who represent the "best and brightest." Through these students' stories, the author shows how "doing school" can interfere with being educated and happy. These teenagers are successful in terms of grades and recognition, but they pay a high price in terms of sleep, peace of mind, ethics, and self-concept. This book is a must read for any parents of high school–age students.

The Homework Myth: Why Our Kids Get Too Much of a Bad Thing, by Alfie Kohn (Da Capo Press, 2006). Alfie Kohn is the author of several books about education and parenting, most notably *Punished by Rewards.* In *The Homework Myth,* Kohn argues that homework is often unnecessary, potentially harmful, and, through middle school, of questionable value. He addresses the issue from the points of view of both parents and teachers, backs up his claims with research, and offers new ways to think about the homework tradition.

Conquering the SAT: How Parents Can Help Teens Overcome the Pressure and Succeed, by Ned Johnson and Emily Warner Eskelsen (Palgrave Macmillan, 2007). Within this book about the SAT is a wealth of good advice for all parents. Standardized testing is in many ways a game, and the authors argue that if we treat it as such, and help our children learn the rules so that they have a good chance to do their best, they can resist attaching their self-concept to test scores. *Conquering the SAT* also offers examples of how to talk to children about school and learning so that they will have a positive view of their own potential and education.

The Underground History of American Education: A School-teacher's Intimate Investigation into the Problem of Modern Schooling, by John Taylor Gatto (Oxford Village Press, 2001), available online at http://www.johntaylorgatto.com. Not for the faint of heart, this book by a former New York State and New York City Teacher of the Year is an extensive look at the history and politics of schooling. Gatto's arguments about the real purpose of school and his thoughts on what constitutes a good education are thought-provoking and, for some people, life changing.

3

HIDDEN TRUTHS OF HOMESCHOOLING

A lot of people think at either two ends of an extreme spectrum: that homeschoolers do the same thing that public schoolers do, only at home, and that homeschoolers just lie around drinking pop and watching TV all day.   — *Orlando*

I constantly am told, "I can't believe you were homeschooled! You're not shy."   — *Tim*

What about homeschooling isn't commonly misunderstood by others? Homeschooling is simply a different approach to learning. Home-schoolers are not all misfits who couldn't do well in school and we are not forced to sit at desks while our parents teach us.... One of the more annoying assumptions is that homeschoolers are not socialized and have trouble meeting people.   — *Tanya*

DESCHOOLING

What if our ideas of education are wrong? What if schooling could be different? What if classroom learning is not the best choice for every child? What if a good education can be easily obtained without school? To help parents answer these questions, many homeschool experts suggest that, before choosing an approach to learning, families first do some deschooling.

"Deschooling" is a word that homeschoolers use for the process of decompression from the effects of school. During deschooling, families take stock of what they know and believe about school, sift through it, question it, keep what is useful, put aside what is not, and make room for new understandings. If families do not take the time to deschool and instead enter homeschooling right where the classroom left off, with daily lessons and tests and days packed with assigned learning, they may burn out quickly and never discover the joys that home education can bring.

When children deschool, they take some time off from formal educational activities. This time off can last several days or several weeks, depending on state homeschool regulations and children's attitudes toward homeschooling. During this time, families can go to parks and museums and nature centers. They can spend leisurely hours in the library or read aloud to each other all morning or talk about what they would learn if they could learn anything they wanted. Deschooling gives children the time and space to rediscover a delight in learning, become reacquainted with their interests, talents, and curiosity, and sometimes simply rest from the emotional and physical work of school.

When parents deschool, they reconsider what a good education really means. We think we understand education because we've all been to school. We have accepted without question that certain subjects need to be studied at certain ages, that learning with a group of age peers is best, and that grades are necessary motivators for learning and are accurate reflections of the quality of learning. We have assumed that the move from the home learning of previous centuries to modern classroom education has been one of constant progress. However, when we look at learning and education from outside the context of school, an amazing thing happens.

Faced with the prospect of homeschooling, we begin to question our previous beliefs. What if subjects can be learned at different ages by different students, depending on their individual strengths and aptitudes? What if learning with a mixed-age group offers more opportunity to share knowledge and less room for unhealthy competitive grade seeking? What if grades are not only unnecessary for learning to occur, but actually inhibit some students' natural motivation, and often reflect neither quantity nor quality of learning?

In essence, what if school and education do not always go together?

Even parents with strong backgrounds in education find that homeschooling opens their eyes to new possibilities for learning:

> *I have to tell you that when I was a public school teacher, I was so limited in my understanding of learning and in what I thought was out there for us to learn from—oh my! I seriously must have thrown out ninety percent of what I used as a teacher.... Homeschooling has opened my eyes to a larger and more interesting world than I ever knew about.*   —Becca

Until we have an idea of what a good education should be, we are unable to decide whether our children would benefit from homeschooling or whether schools are meeting our children's needs.

THE PURPOSE OF EDUCATION

Where can we start to think about education and its purpose? According to John Taylor Gatto, author of *The Underground History of*

American Education, schools in the United States traditionally had three purposes: to help children be good people, be good citizens, and find their particular talents to develop fully.[1]

Most people would agree that these purposes serve as an excellent starting point for a discussion of the goals of education. A child who is learning to be a good person and a good citizen and who is discovering and developing individual talents would certainly seem to be getting a good education. From this starting point, parents can work backward to ask themselves what it means to be a good person, a good citizen, and to have one's talents recognized and developed. Then they can consider what kind of education best meets these ends.

The United Kingdom offers a similar way to think about what it means to be well educated. Parents in England and Wales are required to provide all children of compulsory school age with an "efficient full-time education suitable" to (a) "age, ability and aptitude," and (b) "any special education needs." This education can be obtained through regular school attendance or otherwise.[2] A "suitable education" is defined as one that prepares children for life in modern civilized society and enables them to achieve their full potential, similar to Gatto's three original purposes of American education.

The new perspectives on education that deschooling brings give parents the freedom to make choices. Once we understand that children learn state history in fourth grade because to do it then is convenient for the schools, not necessary for the children, we can make a decision for our own child based on the child's needs and interests, not on someone else's timetable of what every fourth grader should know. Once we realize that grades are an optional part of learning and, for some students, are harmful, we may or may not choose to grade our children's learning. In either case, however, we will never view grades or standardized tests the same way again. Grades and test scores cease to have the same power over a child's (and parent's) self-concept as a learner.

Even for families who do not decide to homeschool, deschooling has many benefits. The parent's role in conventional schooling is confusing at best. How involved is too involved? How much help with homework is enough, or too much? How can parents encourage children to prepare for good colleges without pushing them to unhealthy perfectionism? The rise of the helicopter parent is an indication that we are confused about our roles in adult children's lives, unsure of how much support is too much or too little. We are too easily swayed by the latest parenting guidelines, and we don't know how to hear and trust what we know best: our inner knowledge of our own children.

We can also ask our children their thoughts on education. Even very young children have ideas of what they want to learn and think they should learn, and where they see their education taking them in the future.

> I think that you should know what you need to know, and to be able to do what you want to do.   —Karin, age 14

> A good education is learning how to love to learn. That's something that needs to be with you all your life.   —Tim, age 23

> A good education means to have enough knowledge to support yourself in your chosen field and to be able to maintain a happy and healthy lifestyle.   —Alyssa, age 17

> A good education provides the tools to create good people. A good education not only fills us with knowledge of the world around us, it gives us the ability to seek out more knowledge, the ability to understand what we are given, and the ability to live and learn with other people. A good education produces people who are well

rounded, intellectual, thoughtful people who are not afraid to chal-
lenge themselves to do better. A good education sets us up to be the
best we can be.   *—Orlando, age 18*

A good education is a broad education. One must follow interests as
well as learn the basics needed to move forward in the world.
                                           —Tanya, age 16

BECOMING AN EDUCATED PERSON

Educator and writer Edith Hamilton wrote over fifty years ago
that to be educated is to "be able to be caught up into the world of
thought." She continued, "When I read educational articles it
often seems to me that this important side of the matter, the
purely personal side, is not emphasized enough—the fact that it is
so much more agreeable and interesting to be an educated person
than not."[3]

The great writers, teachers, and thinkers had a very different
sense of education from our current obsession with good grades,
high test scores, and getting into the right college. Edith Hamilton
knew what it felt like to be caught up in the world of thought and to
be a lifelong learner. Her classic 1942 book, *Mythology: Timeless
Tales of Gods and Heroes,* was published when she was seventy-five
years old. Edith learned Latin, Greek, French, and German at home
from her father. She did not attend school until she was sixteen,
and, about the experience, says she and her sisters "weren't taught
anything." She taught herself trigonometry so that she could pass
the entrance exam to attend Bryn Mawr College, where she would
eventually serve as headmistress for twenty-six years. Her writing career did not begin until she was in her fifties, and she went on to
write several books for the general public as well as popular articles
and essays about the classical world and education.

How often do we stop to think about what it means to be educated? How often do we see our children "caught up into the world of thought"? To learn a skill or ponder an idea for its own sake—rather than as a means to a standardized end—seems a luxury for many children today. Parents, children, and teachers alike are so busy chasing objectives, checking off outcomes, improving scores, helping with homework, and filling in gaps that we have lost focus on whether all of this is necessary to be an educated person.

Is education equivalent to passing grades or a diploma? Does it equal a certain number of days in school attendance? Can it be measured with test scores? Does it result in an accepted college application? A good job? Can a person be educated without any of these things? Or does the presence of any of them guarantee that one is educated?

These are difficult questions, in part because we take for granted that we understand what a good education is, and we are not very precise with our terms. We use the words "education," "learning," and "school" to mean vaguely the same thing. We go to school to get an education. Learning happens in school. When we learn enough, we are educated. And so on.

But how much of this is true?

The number one reason that parents give for homeschooling is "better education."[4] While all parents wish for their children to be well educated, families who homeschool must give more than the usual thought to educational goals and what an ideal learning environment would look like. They consider whether education should prepare their children for college, work, or both; to what extent it should nurture habits of thought or develop talents and skills; and what roles good citizenship and moral development will play in daily learning.

These are not simple issues. What it means to be well educated changes with time and place, and even from family to family. A well-educated person from the eighteenth century who traveled in time to the beginning of our twenty-first century would be poorly educated to deal with our electronic, global world. Likewise, were we to drop in on colonial America, we would be poorly educated to handle colonial currency, celestial navigation, and Latin. The first step in making the homeschooling decision is thinking about what a good education is, both in a general guiding sense and in a personal way, for individual children and families in a specific time and place.

For parents whose children go to school, both the educational goals and the learning that takes place during school days to meet those goals are determined by others, either private school officials or local and state governments. If a family's own goals and values conflict with those of the school, parents have only their limited time at home with their children to supplement or offer alternatives to what happens in the classroom. For some parents, this means nightly homework sessions to help children who are struggling. For others, it means using after-school hours to offer enrichment activities for more challenge, or spending weekends learning about themes and subjects important to the family but not addressed in school. For example, Cheryl Fields-Smith in her research on southern black homeschoolers has found that some of the main reasons that many black families give for homeschooling are "infusing black history and perspective into education, countering negative images of black children, [and] demanding higher expectations from their children."[5] These parents have decided what kind of education they want for their children, and they realize that homeschooling gives them the parental involvement necessary to provide this education.

LIFELONG LEARNING

We are born to learn. From the moment we set our eyes on the world, we are growing, developing, becoming educated. Children take joy in learning to walk, to talk, to do things for themselves, not just as means to an end but because the learning is its own reward. Learning is one of life's pleasures! A teenager or an adult who has retained that love of learning for its own sake, who never stops learning because learning feels good, is a joy to see. When the elderly continue a habit of lifelong learning, not only do they gain more pleasure from life, but they may even postpone Alzheimer's disease.[6]

Most parents would say they want their children to be lifelong learners, to be internally motivated to learn. However, we often don't know where to begin or how to overcome the nearly overwhelming influence of grades, tests, academic competition, and other aspects of schooling that, for many children, undermine learning for its own sake. In my college classes, I will sometimes give writing assignments that won't be graded. Some students have no idea what to think of this or how much effort to give. Other students do the bare minimum to get credit for having done the assignment, and see no sense in doing more if there is no grade as a reward. It is sad, because these students are lost without grades; they no longer know how to find intrinsic joy in learning. For them, education is a numbers game.

Homeschooled families know that going to school is not the only way to get an education, nor does it guarantee one. They know the difference between learning to live in society and being socialized. And they are in a unique position to take a step back from the academic competitiveness and hectic schedules of modern life and to see the broader goal of lifelong learning. The rewards of homeschooling are rich and extend far beyond educational milestones of grades or graduation to the heart of what it means to be an educated person.

All families can encourage children toward lifelong learning by urging them to find and pursue their own interests, to trust their curiosity. We can look for the activities, books, and ideas that make our children's eyes sparkle; then, without orchestrating, we can support those interests—even if they are not necessarily our favorite topics—by offering opportunities to learn, ways to practice skills, and plenty of conversation. Allowing children to talk about what interests them is one of the best ways to nurture joyful learning, because it sends the message that their interests are important and worthy of discussion.

Parents often find that homeschooling awakens their own drive to learn, which then enables them to be powerful role models for their children. One of the best kept secrets of home education is that it can be as much fun for the parents as for the children, as families revisit favorite subjects, learn new ways to look at old knowledge, take up new hobbies, and soak up new ideas.

Homeschooling not only nurtures a love of learning, but it offers a chance for children to be around others who love to learn as well. Homeschool support groups and learning groups are often filled with children and teens who display an unrestrained curiosity and thirst for knowledge. They know they will not be graded and they are not in competition with their friends, so they are free to learn for the joy of it. It doesn't take long for a child new to homeschooling to realize that this is a place where it's okay to love knowledge.

Interestingly, while we put high value on internal motivation, our expectations for children are low. We don't expect children to love learning for its own sake or to exhibit traits of lifelong learners. Many of the quirks or oddities noted in even positive profiles of homeschooled children are, in fact, indications of self-directed minds. Consider a *Time* magazine article about Christopher Paolini, who wrote the best-selling novel *Eragon* when he was fifteen: "One expects home-schooled kids to be a little odd, and Paolini is—just a little."[7] What makes him odd? According to

Time, he looks a little like Harry Potter. He is "overarticulate," and his hobbies include the crafting of medieval weaponry. We just aren't used to seeing children who are caught up in the world of learning or who show a passion for the world around them.

THE MISUNDERSTOOD HOMESCHOOLER

Homeschoolers like Christopher Paolini are widely misunderstood and are expected to be "odd." The stereotype lingers of the home-schooled child who has no friends, cannot make a successful transition to group learning environments, has a sheltered and unrealistic view of the world outside his home, and either isn't keeping up academically with peers in school or is a child genius. Homeschooling parents don't fare much better, portrayed as overprotective or pushy, incapable of the task of educating their children, or antidemocratic for their refusal to accept the government schooling available to them.

Many of the myths—for example, that children who don't go to school have no social skills or that they have a hard time getting into college—have been successfully addressed and corrected in the national media.[8] However, some people don't seem to be paying much attention. Homeschoolers still hear many of the same questions. One homeschooler said the most common misconceptions she hears are that "either we are super smart and the next Einstein, or we just get to sleep all day and eat ice cream." Of course, the two stereotypes are not mutually exclusive.

These stereotypes can make prospective homeschoolers think twice about doing something so seemingly different from the norm. After all, what will the neighbors think? Let's look at two of the most common myths—the Socialization Myth and the Genius Myth—to see what they really mean, and the hidden truth about homeschooling that each one contains.

The Socialization Myth

Socialization is homeschooling's proverbial dead horse. Dozens of articles and chapters, conference seminars, and at least two books have focused specifically on the topic of the socialization of home-schoolers and debunked the concern. So why is the question still being asked?

Whenever I am asked "What about socialization?" I think of a homeschool literature and writing group I have worked with for many years. Approximately fifteen children, ages ten through eight-een, gather weekly to discuss novels, plays, poetry, and nonfiction, and to practice and share their own writing. The children vary not only in age, but also in their personalities, homeschooling styles, and reasons for homeschooling. Every week they read aloud, share opinions, offer criticism, and learn from each other. Often they work in small groups of mixed ages, and the groups instinctively offer each member a chance to contribute according to his or her talents and comfort level. If one child struggles or needs reassurance, another child almost always steps in to help. They also are able to work alongside parents who sometimes sat in on the group.

These children aren't perfect, but they have social skills that most adults would envy. They listen to others, offer honest yet tact-ful feedback, support each other as learners, and work together as peers regardless of age or ability. Of course, many children who go to school also have these skills. Homeschooling, however, makes them easier to practice and to acquire.

When people ask about socialization, they are usually asking about something more specific: Do homeschooled children have any friends? Do they spend a lot of time alone? Do they know how to meet new people or interact in a group? Sometimes the question really is this: Can the homeschooled child conform and follow the rules?

An article in the *New York Times* posed the socialization question in a way that gets to the heart of the misconception: "Will the [homeschooled] children interact enough with peers to build communication skills and develop emotionally?"[9] Implied in this question is that extensive interaction with age peers is necessary for building communication skills and developing emotionally. We can turn the question around and ask, "Can children in school have too much interaction with peers, impeding their emotional and social development?"

While being with other children of a similar age is certainly not a bad thing, schools institutionalize same-age socializing and learning. Children in school have no choice but to spend at least six hours a day with twenty or more students of the same age. These groups are further separated into elementary schools, middle schools, and high schools. This model is such an engrained part of our educational system that it is hard to imagine anything different or to see the problems it can cause.

Sadly, some children are indeed isolated, feel friendless, or have trouble finding their place in a social group; however, these children can be found everywhere—in public schools, private schools, and homeschools. Feeling that we are a part of society—that we belong—is a human issue, not a homeschool issue. If we think back to our days in school, we will remember times when we had trouble making or keeping friends and we can think of others who were known as loners. When it comes to friendship, there is no guarantee that a child will find a friend in a group of age peers. For some children, being with age peers makes it less likely they will find friends.

Friendships are based not on age but on interests, personality, values, and goals. Adults can think about their own friendships. How many of your friends are exactly your age? What is the basis for your friendships? Where did you find most of your friends? For children more than adults, individual development has a great effect on friendships. Children who are at a higher or lower level of emotional development than most other children their age may have a hard time relating to other children in their school grade. Children who have unusual interests or hobbies may need to look outside school to find activities where they can meet others who share their passions.

Finding friends is not a problem for homeschoolers. First, they often form real friendships within their families. Siblings and extended family members have the time to get to know each and learn to live together, if not always in perfect harmony, at least with a certain degree of understanding. Homeschool parents often find that it really is possible to be friends with their children while also exerting adequate parental control and guidance. Kathleen, who has homeschooled five children, says one of her greatest joys is "lingering over tea after breakfast with delightful teenagers who are, at the same time, my best friends and my children."

Homeschooled children find friends in home-education support groups and learning co-ops, sports or fine arts classes, and church and community organizations. Many homeschooled children also form or keep friendships with children who go to school. Tanya sums up how many homeschoolers feel about friendships and socialization:

> *I have yet to meet or even hear of a homeschooler who had social problems or trouble meeting people simply because of homeschooling. I personally attend a literature and writing class along with sixteen of my friends. I also go to a homeschool group every other week where we learn something new and interesting (how to tie-dye or a session on molecules) for two hours, then for two hours we can just hang out and have fun. I also have a couple friends from the local high school. There are many other different groups a homeschooler can join.*

In a series on middle schools, the *New York Times* explored some of the drawbacks of limiting children's exposure to younger and older students. By removing sixth, seventh, and eighth graders from the more nurturing approach of elementary school, we may be asking them to make too abrupt a transition at a time when they are experiencing important physical and emotional changes. The most popular proposed solution—schools that include kindergarten through eighth grade—is criticized because it can hold the older students back from the academic preparation needed for college. Some schools are experimenting with combining sixth through twelfth grades, to give young teenagers the head start they need on college-preparatory learning and provide them with the example of older students. However, this option can force children to have to deal too quickly with high school peer pressures without the opportunities to serve as role models and leaders themselves.[10]

Balancing academic challenge with social and emotional stability is less of a problem for homeschoolers. In families and in homeschooling groups, homeschooled children have the opportunity both to learn from older students and to be models for younger children. Much like one-room schoolhouses, home-schools give students the freedom to find their own level of learning and to move as quickly or as slowly as their needs dictate. Homeschooling parents often teach the same subject matter to children of different ages, with the confidence that each will learn according to need and ability.

Mixed-age learning also allows children to find learning peers of similar ability for specific subjects, regardless of age or grade. In a homeschooling group, it's not uncommon to see groups of younger and older children playing chess together, discussing books, or helping each other with math. In schools, grouping by ability rather than age is controversial and is hard to implement. In homeschooling, it is a natural part of learning.

But what about socialization as conformity or fitting in? The implication here is that children need to be able to get with the program, go along with the crowd, or be a team player. And this is where the Socialization Myth reveals a hidden homeschooling truth: home-schooling does not nurture this kind of socialization nearly as well as schools do.

The social education that homeschooling can offer is something very different. Linda Silverman, an advocate for gifted children, re-sists the concept of socialization and prefers to speak of social devel-opment, a "deep, comfortable level of self-acceptance that leads to true friendships with others."[11] Social development is not depen-dent on whether we are accepted by others, but on how much we learn to accept ourselves, which then enables us to reach out to oth-ers in friendship. Acceptance of self begins with and is nurtured by the family.

Homeschooling's social education is closer to sociology's con-cept of socialization than to fitting in or being popular:

> *Socialization* is the process whereby, through contact with other
> human beings, the helpless infant gradually becomes a self-aware,
> knowledgeable human being, skilled in the ways of the given cul-
> ture and environment.[12]

It soon becomes clear that people who ask the socialization question of homeschoolers are really not asking about socialization at all, because the process of becoming "a self-aware, knowledgeable human being, skilled in the ways of the given culture and environ-ment" starts and continues in the home. Contact with the human beings who know and love us best, our parents and family, helps us to know and value ourselves, and, eventually, to learn how to take our place in the adult world.

One clear advantage that homeschoolers have is the ability to interact with adults:

I think homeschooling is the reason I so easily interact with people from many different sorts of backgrounds. As a kid I was able to converse with people much younger and much older than I. I did so on a regular basis and more so than my schooled peers. I was not limited in my friendship to my grade; instead, it was limited only by intellect.   —Tim

Often when I go out into situations where I interact with adults— when we have guests for example—they always remark on how well I am able to talk with them on their level. In my opinion, this is because homeschoolers are around adults in social situations more often. Instead of just teachers, homeschoolers interact with lots of other people—other homeschool parents, guest speakers at homeschool events, and much more.   —Tanya

One of the most frequent comments I get is "Wow, you are really social for a homeschooler!" or something to that effect. I have found that because I'm homeschooled it is easier for me to communicate with all ages and not only my peers. Many kids my age are put into an age exclusive environment, whereas as a homeschooler, I interact with all ages, making it easier for me to be social with not only my peers but all age ranges.   —Yasmin

Homeschoolers learn to get along with all age groups, because their learning peers include both the young and the old.

Some homeschooling families find that their pursuit of adequate socialization and group learning leads to the very stress-filled, hectic schedules they were trying to avoid. Out of a desire to take advantage of the many outside-the-home opportunities available or to deflect or pre-empt socialization criticisms, homeschoolers can easily find themselves spending much of their time in the car as they juggle activities such as music lessons, sports practices, and museum classes. Manfred Smith, president of the Maryland Home Education

Association, comments that the problem for many homeschoolers is not lack of socializing, but too much group learning that can get in the way of the individual learning that is unique to homeschooling. "Parents are rushing around to make sure their child has ninety-two opportunities. They are in this class, that dance thing; on the weekends, there is a lot of soccer," Smith says. "People have begun to back off on the socialization."[13]

Anyone homeschooling or thinking of homeschooling can remember that one of the unique benefits of homeschooling that money can't buy but that can easily be lost in pursuit of too much socialization is time: time to work slowly, time to think, time to focus on a single idea rather than multi-tasking, time for self-knowledge and reflection, time to question the world and wait for the answers. Karin sees the value of protecting a certain amount of unscheduled time when she can learn uninterrupted: "The social life is really not a problem. There are only two days a week that I don't have something that happens on a regular basis, and I actually like those days because I get to be home without thinking about when I have to leave."

Rather than ask, "What about socialization?" parents who are considering homeschooling can think about how they will provide a social education for their children. What are the goals of social education, and how would we know if those goals are reached? Does classroom education meet these goals? Does socialization mean conformity? If not, how can we teach children to follow their own inner voice while at the same time working well with others? Are parents and children willing to seek out other homeschoolers for support and friendship? What community, church, or other activities will the children be involved in?

Another important consideration is how a child's temperament affects socializing. Are the children introverted learners who need more than the usual amount of time alone, or are they extroverted types who will need a wider variety of social opportunities? If a parent is introverted and the child is extroverted, homeschooling can be a challenge, because the child may thrive on a level of socializing that leaves the parent exhausted. In this case, parents can look for some homeschool co-ops and group learning experiences that do not require full-time parent involvement.

Likewise, an extroverted parent may assume that children will quickly become bored and restless if they stay home much of the time. However, homeschooling works particularly well for many introverted children precisely because it allows for more time at home, away from large groups. The social education of an extremely introverted or shy child may begin with family or small groups of two or three children. Later, as the child's comfort level and skills increase, groups can get a bit bigger. However, even older introverted children can do very well on what others would consider a meager social diet. Quality of social interaction is often more important than quantity.

I think homeschooling can be an excellent way to raise socially balanced children. A naturally shy child can be gradually accustomed to a larger social circle of friendly adults and peers, while a naturally gregarious child can find an outlet for his needs. Homeschooling parents have much more control of social situations and can adapt them to their individual child's needs if they need to make it a priority.
—Kathleen

Children who learn at home gain a different set of social skills from many other children. Homeschooled children are excellent at talking to and working with children of different ages. They have interesting and engaging conversations with adults. They are also unlikely to follow the ideas of a group without first making up their own minds. College students who were homeschooled have told me that they feel more mature than their dorm mates, because they know how to think for themselves and aren't influenced as readily by peer pressure.

Becca, like many homeschoolers, thinks that the question of socialization is a non-issue: "I really don't give any thought to 'socialization'; we just enjoy relationships with other folks." Debbie agrees: "My kids feel very comfortable in all kinds of social situations. They can carry on conversations with all kinds of people. I am not concerned with my children's socialization at all."

The Genius Myth

> *Non-homeschoolers think one of two things about us: that we are really smart "geniuses" or that we cannot possibly be as smart or well prepared as public school kids. Neither of these is true, obviously.*
>
> —Orlando

You've probably read newspaper or magazine articles about homeschooling wonder kids. They win spelling bees, go to Harvard when they are sixteen, or get perfect scores on the SAT. In a well-intentioned effort to legitimize and defend homeschooling, some researchers have tried to prove that homeschooled kids are not just as smart as other children, but smarter. They don't just keep up with their age peers in school, they are ahead of them.

Just as with statistics about the numbers of homeschoolers, claims about homeschoolers' intelligence as a group or their educational achievements are questionable at best. Many homeschooling families opt out of standardized testing. Others don't use grade levels. Research studies often focus on specific segments of the homeschooling population that are not representative of all homeschoolers. Profiles of homeschoolers tend to focus on children with unusual accomplishments because they are interesting, not because they are the homeschooling norm.

What this media attention has done is show that homeschooling can meet the academic needs of children, and that children who homeschool can do well on tests, go to college, and succeed according to societal standards. This is a good thing. However, these reports may lead some parents to wonder not if they as parents are qualified to homeschool, but if their kids are smart enough to homeschool. Or they may expect that homeschooling will automatically make their children more intelligent and more accomplished.

The hidden homeschooling truth in the Genius Myth is this: homeschooling can give all children the chance to develop their potential and gifts fully.

I will admit that I find homeschoolers to be exceptionally curious and bright. Why wouldn't they be? They are able to get enough sleep to be healthy and awake. They don't confuse grades or class rank with learning. They can explore their interests and use their strengths. They have the freedom to learn as much as they want, even if it's above their grade level or a subject that isn't covered in schools. Like toddlers, they have the chance to approach the world with wonder and eagerness, ready to take the next step and to see where it leads. It should come as no surprise that children who otherwise might not test well or be star classroom pupils would be able to find their unique gifts as homeschoolers.

Unreasonable classroom expectations are often a reason that parents turn to homeschooling:

> There was an initial dissatisfaction with how our son's first grade teacher and school were putting a lot of pressure on him to read at a certain level. After some time, this child who loved books began saying that he hated books. He began feeling set apart in a way from his peers. —Debbie

As a homeschooler, Debbie's son has been able to regain his love of books and all learning. He is curious and bright and is quick to share his love of books and his own writing with the writing group he attends every week. No longer set apart, he now feels like a member of a learning community.

Children who are precocious learners often thrive in homeschooling. Home education can offer gifted children a learning environment with few constrictions on how much or how deeply they learn, and a social and emotional environment that respects their innate sensitivities and intensities. Homeschooling is fast becoming a preferred educational option for many gifted children, especially highly gifted children whose intellectual abilities cause them to be far out of step with grade-based curricula and expectations.

DEALING WITH CRITICISM

In remarks at a home-education conference, one homeschool parent described the reactions he received when he decided to homeschool his two children:

> When we began homeschooling almost ten years ago, it provoked vague unease in most people. The most common reaction I got was a quick change of subject, as if I'd just admitted I had six toes. Or, if someone did have any more to say it was the infamous "What about socialization?" or comments such as, "Well, and when will your children enter school?" In other words, homeschooling was viewed as an aberration and, since I seemed to be an otherwise sensible person, it was assumed I would eventually wake up, conform to the norm and put my kids back where they belong.[14]

Parents who homeschool learn to get used to such reactions. In time, as people get to know our children and see how homeschool works for us, they usually understand better and may even begin to wonder if they should consider homeschooling. At the

beginning, however, negative reactions or well-intentioned suggestions that perhaps we don't know what we're doing can be difficult. A veteran homeschooling parent offered this good advice to
new homeschoolers:

> *I think the criticisms often result from misunderstandings and simple*
> *communication usually takes care of it. I have also noticed that even*
> *skeptics often admire homeschooling families when they see how well*
> *the children are doing and what nice people they are. So, ultimately, I*
> *think the best strategy is to live a good life and be true to your own*
> *values. Of course, there are always those few individuals who will*
> *criticize, no matter what. I try not to pay too much attention to that.*
> —*Susan*

My own experience is that homeschooled children are friendly,
open, and generally easy to talk to. They are as reserved or as out-
going as anyone else, have a boatload of friends or a few close bud-
dies. They may get up early, sleep until noon, dress as they please.
They can be wonderfully average or amazingly bright, and some
struggle with conventional learning tasks but excel in more unconventional areas.

The hidden truth is simple: homeschoolers are not all that different from everyone else.

4

HOMESCHOOL DAYS

What Does Learning at Home Look Like?

Ben wakes up to his alarm every morning at 6:00 A.M. He has a reassuring routine that is rarely broken: shower, dress, breakfast, math (he does math most easily when he is fresh in the morning), history, outside play with his siblings, lunch, science, literature and writing, family chores, supper, free reading, family time, bed. To fulfill his state home-education requirements, he takes a standardized test every year, and his parents keep a detailed portfolio of his work and progress. All of his textbooks and other materials come from one company that specializes in homeschool curricula. Ben is graded on weekly tests in most subjects and receives quarterly report cards from his parents.

Emily has recently been learning everything she can about her hometown of Chicago. She and her family have put together a course of study that includes documentaries on Chicago from their library; frequent trips to the Art Institute where she is learning about Impressionism; architectural walking tours of her city, the birthplace of the skyscraper; learning about the World's Columbian

Exposition of 1893; taking in advances in science displayed at the Museum of Science and Industry; and reading the poetry of Carl Sandburg. Emily keeps a scrapbook of what she learns and is writing her own history of the city. She is not graded on what she learns, and her learning does not correspond to one specific grade level.

Evan is currently enrolled in four distance-learning high school courses. Two of these are online courses; for the other two he uses textbooks and mails his assignments. His mother works outside the home, and his father is self-employed in the home. Evan usually works side by side with his father. He asks questions when he needs to, but most of the time he is responsible for the pace and completion of his studies. Online tutors are available with two of his courses, and Evan frequently emails his tutors with assignments or questions. In the late afternoon, Evan volunteers at a youth center, and on the weekends he is on a homeschool soccer team.

Kendra wakes up when she is rested, sometimes very early—even before her parents—and sometimes mid-morning. Likewise, she may go to bed soon after supper if she is unusually tired, or stay up until midnight. Her days are never the same. Some days she reads almost nonstop: fiction, nonfiction, poetry, plays. An avid reader since age four, she keeps stacks of library books in her room. Other days she writes short stories, or practices piano for hours at a time, or spends time at a local nature conservatory. Kendra is a self-directed learner who follows her interests and curiosity. Her parents do not use any testing for her, and the state they live in does not require it. She was recently accepted into a summer intern program at the city zoo, and she hopes to attend college full time when she is seventeen.

Which of these children are homeschooled? If this were a multiple-choice test, the answer would be all of the above. Ben is a traditional homeschooler whose education is similar to that of many schools.

Emily's unit study on Chicago allows her to cover several subject areas through one main topic. Evan uses distance learning for much of his homeschooling. And Kendra learns according to her own motivation and interests. While their approaches to education are very different, they have one thing in common: They receive their education outside of conventional public or private schools and under the guidance of their families. Just as all public school students cannot be lumped together in a representative group, so, too, homeschoolers are too diverse to be understood in generalizations. Families have different priorities and ways of life. Children have different ways they learn best.

Curious minds, however, still want to know: What does homeschooling look like? What do homeschoolers do all day? How do they learn?

HOMESCHOOLERS IN PAJAMAS

"Do you get to stay in your pajamas all day?" Homeschoolers either love or hate this question. Those who love it say, "Yes! I can learn in any clothes I want." Those who hate it feel that the person asking the question is really asking, "Do you take your learning seriously?"

Personally, I find the pajamas question amusing. Our son doesn't stay in his pajamas all day, but there are many days when he dresses quite comfortably, in shorts and a t-shirt, socks and slippers. Homeschoolers are literally at home in their learning environment. When we get together with other homeschoolers for group learning activities, the children usually arrange themselves in a circle, all with their shoes off, some on chairs and others on the floor. Young children sometimes sit on their mothers' laps. Teens stretch out as much as they can. They are not in their pajamas, but neither are they in uniforms or dress shoes. Think of children who do homework after school or on weekends. They change into comfortable clothes, arrange a nook where they can sit comfortably, then get to work. It's the same for homeschoolers, just for longer time periods.

The companion to this question is, "Do you get to sleep as late as you want?" Again, some homeschoolers respond with a grin, "Sure! Then I stay up until after midnight reading." Others feel they are being asked to justify their freedom from school schedules and to prove that they are not lazy bums who are wasting their youth.

The hidden homeschool truth in these questions is that homeschooling provides almost unimaginable freedom and flexibility in both learning and family life. Homeschoolers have the freedom to discover and honor their own learning styles and biological rhythms rather than try to make themselves fit into someone else's model of where and when to learn. Families have the flexibility to schedule their lives in harmony with physical and emotional needs rather than sticking to a timetable that keeps everyone moving in the same direction at the same hurried pace.

Debbie, whose children were in school before being homeschooled, says that one reason she likes homeschooling is that it allows her children to get enough sleep. One of the undervalued benefits of the freedom and flexibility of homeschooling is sleep. When children get the sleep they need, they learn better, they are healthier, and, as all parents know, they are more pleasant to be around. According to *Science News Online,* by sixth grade many children have begun a pattern of sleep deprivation that continues through their teenage years.[1] Because teens require even more sleep than children or adults, the problem of sleep deprivation worsens in high school. Also, many adolescents naturally experience a shift in circadian rhythms, causing them to stay awake at night longer than they had before.[2] When this is combined with high school starting times of 7:30 A.M. or earlier, teenagers have a hard time getting the sleep their bodies and minds require.

Parents benefit from homeschooling's freedom and flexibility as much as children. With no need to rush to meet morning buses, keep track of lunch bags, juggle two or more school schedules, review and sign homework forms, and try to keep abreast of what goes on during the school day, they have more time and energy to devote to the important aspects of parenting. Many parents can't imagine themselves homeschooling because they think it would take too much time; however, they forget about the time they already devote to their children's school needs and schedules. Homeschoolers also find that the job of parenting is easier because they know their children better and have the time to discuss and deal with problems as they arise.

When families no longer have to accommodate bus schedules and school bells, their days become their own, free to schedule in ways that work for them. For some families, not much is changed: they use alarms to get up early, start the learning day after breakfast, and end their study in the afternoon. In other families, children might sleep until they wake up naturally and extend their learning into late evening. Or they could integrate their education in all that they do and not worry about schedules that distinguish learning from nonlearning times.

ISN'T HOMESCHOOLING JUST SCHOOL AT HOME?

When I was young, one of my favorite pastimes was to "play school." I set up a chalkboard, papers, pencils, and books, and forced my brothers and any other children who might visit to sit in chairs and be the pupils to my teacher. While the insistence I then displayed now makes me cringe, the feeling of being in an atmosphere of learning and an environment rich with the symbols and tools of learning was almost magical. Doing school at home was not stifling; it was fun.

Many people might assume that what goes on in home education is much like playing school. Parents take on the roles of teacher and principal, and the children are dutiful students. The basic objectives, techniques, and resources remain the same as for conventional school; all that is different is that the four walls that make up the educational setting are that of a home and not a school building. In reality, however, home education often bears little resemblance to school, as the examples at the beginning of this chapter show.

The School at Home Debate

Nothing can spark a debate more quickly in a homeschool group than a discussion of parents who do "school at home." For some homeschoolers, this catch phrase is used in a negative way to describe institutionalized classroom education in the home, complete with chalkboards and recess bells and raising one's hand to use the restroom—similar to my own playing school as a little girl.

In reality, few homeschoolers would describe themselves as doing school at home in the sense of trying to replicate classroom education exactly. Certainly many families do use purchased curricula for all subjects. Others follow a detailed daily schedule or encourage self-discipline or consider the parents to be teachers in a formal sense. However, by the very nature of their placing education within the realm of the family and community and outside the conventional classroom, their homeschool experience is more than just a mirror of government or private schools. Even homeschool parents who use textbooks and follow a daily schedule of classes and use grades provide advantages of personal attention, individualized lessons and goals, and educational flexibility not available in large classrooms.

Another important difference is that the time spent on academic activities is, for homeschoolers, usually a fraction of the time spent in school. Imagine (or remember) a typical school day. How much time is spent going from classroom to classroom, taking attendance, getting out books, making sure everyone in class is ready for discussion, waiting for over twenty students to have a chance to respond to a question, reviewing material that some students already know and others aren't understanding because they need individual help? Lacking such delays and distractions, homeschoolers easily accomplish in two or three hours of focused learning what takes an entire day at school. For some families, using traditional learning methods and materials for these few hours is an efficient way to cover state-mandated curricula or to give child and parent a positive sense of accomplishment.

Classroom education is what most adults know best, so it's natural for some families to use this model when they start homeschooling. Also, children who come to homeschooling after one or more years in the classroom often are comfortable with the ideas of grade levels, tests, discrete subjects of study, academic periods, and textbooks. Some homeschool families start with these familiar concepts and then modify or move away from them as they deschool and become more aware of and secure in homeschooling freedoms. Other families find that a routine similar to school suits them well, and they follow the same academic calendar and curriculum as their local schools. And some homeschoolers don't have a choice: their state laws require them to use school-approved curricula for each grade level and to keep attendance records during academic terms.

Does this traditional school-inspired approach work in home education? Not for everyone, especially those who choose homeschooling expressly to get away from the trappings of the classroom. Homeschooling is all about freedom: the freedom to learn about history all day rather than for fifty minutes, the freedom to learn according to curiosity and interests rather than grade-level lists of topics, the freedom to zoom ahead in areas of strength and take as much time as necessary to learn difficult subjects.

Our own homeschooling starting point bore little resemblance to school. A freer and more organic form of learning suited my own temperament and the needs of our son. His innate curiosity was strong, and I was happy with what he learned as he followed his own interests and was allowed to ask plenty of questions. Our days were never the same and would unfold according to whatever our son was interested in. We homeschooled year-round, with no appreciable difference between a day in January and a day in July.

Then, when he approached high school, he requested a more school-like schedule, and we modified our homeschooling to meet his needs. We followed an academic year for most subjects—but not all—and he enjoyed having summers relatively free. We paid closer attention to individual subjects, distinguishing between what he had already learned and what he needed to learn for college preparation. We researched physics and calculus textbooks, and he asked me to write syllabi for both subjects. He also asked me to take on more of a teacher role than I had in the past.

Our son did not lose his strong sense of self-direction nor his love of learning during this time—in fact, both continued to grow—and our commitment to homeschooling never wavered. While it may have looked from the outside as though we were doing school at home, it certainly didn't feel that way. Our homeschooling didn't feel that much different from what we had done before, perhaps because we were just continuing to meet our son's needs.

Very few homeschoolers desire to make their homes a school. That would defeat their reasons for homeschooling! For some families, however, the familiar routines of school offer a good beginning for home education. Perhaps they do not feel confident enough to branch out far on their own, or perhaps their children thrive on a degree of structure and routine that would make other children frustrated. In any case, dismissing these families as "school at home" homeschoolers is neither accurate nor fair, and homeschoolers are wise to avoid such divisive labels.

What unites homeschoolers is not their methods, but their exercising the right to homeschool. The Wisconsin Parents Association (WPA) urges homeschoolers to take the longer, wider view when these kinds of divisions arise, and to focus on similarities rather than differences:

> We believe there is no one right answer for everyone. Each family must decide for itself what is most important and works best when it comes to the specifics of homeschooling. One of our objections to conventional schools is that they do not allow enough flexibility to meet the widely varying needs of different individuals.[3]

Doing School: A Caution

Although there is nothing wrong with borrowing techniques and tools from the classroom, "doing school" in the sense of going through the motions of a school day without thought or purpose does not result in a good and rewarding education, whether at school or at home. Denise Clark Pope argues that students today are so busy doing school that they miss any real engagement with learning, and they (and we) lose sight of the fact that becoming educated is what school is all about.[4] Another reason for new homeschoolers to take the time to deschool is so they won't continue to confuse the motions of doing school—filling out worksheets, taking standardized tests, following blindly a generic course of study—with the process of learning.

TRUSTING A CHILD'S INNER DRIVE TO LEARN

Our early homeschooling can be described as self-directed learning, based on our child's interests, questions, and drives. Young children naturally resist separating education into individual learning times and subject areas: geometry, science, reading, and so on. Instead, they follow an interest in any and all directions it might lead, for as long as the interest and questions continue, regardless of whether the subject matter is normally covered at a certain age or grade. Many people refer to this kind of learning as unschooling or, in Europe, "natural learning."

Self-directed learning might seem a strange idea at first because the emphasis is not on teaching, but it really is nothing more than learning for its own sake. People who continue lifelong self-education do this all the time. For example, adults are self-directed learners when they teach themselves a new foreign language—just for the fun of it—or learn new cooking skills or take up a new sport. School children whose curiosity and drive to learn have not been thwarted are self-directed learners in the summer as they bury themselves in library books or write poetry or learn a new computer program.

The best way to understand how self-directed learning works is to see it in action through the words of the learners themselves:

> *Instead of sitting at a desk for eight hours a day learning about something we hate, we get to learn about what we want, when we want to. . . . I find I learn things faster than I would in school because I can go at my own pace. When I get interested in something, then I learn about it, and I can skip over the parts that are easy for me and spend more time on the hard things. [My long-term goals are] I would like to improve my playing ability on all the instruments that I play, so I am practicing and taking lessons. I would like to be able to speak French fluently, so I am taking lessons from a computer course. I*

would also like to eat home grown watermelon, so I will be planting
seeds soon. —Karin

I wouldn't say that I have a "typical" day of learning, but recently I
have been playing a lot of guitar and singing, lots of music all
around. I have also been working on photography, both digital and
film. I recently started to work with moving old records onto the com-
puter and, after cleaning them up, burning them onto CDs. I have
also always been interested in string theory and science in general.
—Alyssa

Families do not have to subscribe to an unschooling ideology
nor do they even need to homeschool to enjoy the benefits of child-
led learning. Now a teenager, our son takes a distance-learning class,
a college class, and learns math and science from textbooks, yet he
still gives himself what he calls "unschool days" when he takes pleasure in free reading, looking up answers to persistent questions, and
simply exploring the ideas and objects in his world. Other families I
know integrate aspects of self-directed learning by emphasizing the
delight of learning for its own sake, refusing to use grades and evaluation unless necessary, taking stock of and building up the child's interests and passions, and, going back to John Holt's theme, trusting
the child's ability to learn.

MINDING THE GAP

Critics of homeschooling sometimes question whether children will
have gaps in knowledge or if subject areas may be missed altogether.
They wonder what happens when children simply don't want to
learn anything, and ask how self-directed learners prepare themselves for college and the workforce. The idea of learning gaps is particularly interesting. With today's vast stores of knowledge and ease

of finding information, preventing "gaps" in learning seems less important than choosing what to learn and knowing how to learn it. If any child—homeschooled or not—pursued the goal of a gap-free education, the student would be continually working to catch up, and much of the joy of becoming a well-educated person would be lost. This is exactly what happens in many schools today.

Learning gaps come in two forms: real gaps that prevent us from moving on—from doing what we need and want to do—and gaps that need to be filled in primarily to do well on standardized assessments. Real gaps can usually be addressed quite easily. For example, suppose a student hasn't learned sufficient math to prepare her for college-level work. She has several choices. If doing college math is important to her, she will be motivated to learn the math she needs, perhaps over the summer, or with a tutor, or through an online course. Most colleges offer courses specifically for students who need remedial or refresher courses in math before moving on to more advanced work. That's why the courses are offered: to help students address gaps before moving on. Gaps are expected, to a certain extent. Real learning gaps show us what we need to learn.

Artificial gaps, on the other hand, are a problem only so far as they affect test scores. Artificial gaps are educational hoops, and whether we need to jump through them depends on individual goals. If the student above decides she wants to attend a different college that does not require the SAT and her homeschool portfolio is strong enough for admission, or if she decides to forgo college altogether and take an apprenticeship or study at a two-year vocational school, the gap is no longer an issue, because it does not prevent her from continuing her lifelong education or meeting her goals.

Ken Danforth, executive director of a learning center for homeschooled teenagers, reminds us that we all have gaps in learning:

Most kids who go to school don't learn everything that is offered, or retain it. How many students in school even make the honor roll? Twenty-five percent? Schooling is no protection against learning gaps. What we want are people who know what they know and are honest about what they don't know. People who are willing and able to learn things they need for the next step. The home-schoolers I know aim to be strong in their passions and deal with their "gaps" as needed. Sort of like adults![5]

As an adult—and especially as a homeschool parent—I am confronted with my own gaps every day. I remember some aspects of history more than others and retain only some math concepts that have been useful in my daily life. My own education didn't teach me everything. In high school, for example, I dropped out of chemistry to be the piano accompanist for the choir. When our son was ready to learn chemistry, I found myself faced with a chemistry gap. However, home education has helped me to see these gaps not as signs of failure but as opportunities to learn.

What about times when children don't want to learn anything? Advocates of child-led learning would say that children's natural imperative is to learn, and, unless this drive is harmed or thwarted, curiosity will motivate the child to continue to learn. Even a lazy afternoon of daydreaming can result in hours of creative thinking and problem solving. And when a child starts thinking about college, he or she will have all the motivation necessary to learn what is needed to meet individual goals.

That said, it is important to note that child-directed learning is not a surrender of parenting or a chance for parents to ignore the responsibility they have for their children's life education. All children—homeschooled or not—can lack self-discipline or parental guidance and example. If a child ceases to show interest in or engagement with life, parents would be wise to look carefully at what might need to be changed in the child's environment, habits of thought, self-concept, or family dynamics. Once again, this is more about what makes a healthy and involved family, which in turn makes it easier to have a successful homeschooling experience.

When child-directed learning works well, it can require even more parental involvement, time, guidance, and skills than more traditional homeschooling approaches. This is because a child's curiosity and questions are endless and resistant to lesson plans or any one-size-fits-all approach to learning. Homeschooling parents know how to keep on their toes, if only to reach the highest row of library books!

OTHER WAYS TO HOMESCHOOL

In addition to self-directed learning and a conventional approach to education, homeschool families use a wide variety of ways to learn. A few of the most popular approaches include using unit studies, classical homeschooling, and distance education and e-learning.

Unit Studies

Unit studies allow children to learn about one topic or interest through a variety of subject areas. Unit studies can be developed by parents and children to meet specific needs, or they can be purchased by companies that specialize in finding interesting ways to study popular topics. Some homeschool families do most of their learning with unit studies; others use unit studies for a portion of their learning. Unit studies are often literature based, and they also can tap into a child's innate motivation to learn.

When one of my brothers was young, he went through a spider phase: all he wanted to think about was spiders. We were not a homeschooling family, but he approached his spider self-study with homeschooling drive and enthusiasm. He read everything he could about spiders, learned their anatomical names, and talked about spiders constantly. As a self-directed learner, he learned quite a lot.

Homeschool parents might use these kinds of interests and passions to facilitate a more organized unit study for their children and widen and broaden the learning that takes place. Using the spider example, the child could read *Charlotte's Web* (or the parent could read it aloud), learn about spiders' role in mythology and Native American folklore, examine the geometry of spider webs, watch a science documentary about arachnids, and study species of spiders from different continents and countries. A quick Web search yields a wide variety of free resources to extend and organize the study further, from online videos to complete unit studies that integrate math, science, and language. For older students, the International Society for Arachnology Web page provides stores of information and additional links to learning resources.[6]

Parents often design these thematic studies on their own, either as a formal course of study or just while answering children's questions and following their interests. Published unit studies are also available, either free on the Internet or for sale, as are guides for homeschoolers and teachers for writing their own theme-based curricula. Teacher supply stores carry many thematic studies designed for the classroom but easily adapted by homeschoolers.

Classical Homeschooling

Homeschooling saw a growth of interest in classical education with the publication of *The Well-Trained Mind: A Guide to Classical Education at Home,* by Jessie Wise and Susan Wise Bauer. Classical education focuses on critical thinking and the mastery and use of language, and divides a child's elementary and high school years into stages: the early grammar stage, roughly through grade four or five, when children learn about the rules and facts of the world; the middle logic phase, typically middle school years, that focuses on cause and effect relationships, abstract thought, and the logic of math and argument; and the final rhetoric stage of high school, when the student builds upon and uses the previous stages for self-expression and specialization.[7]

Classical education, which can be traced back to the Romans and Greeks, was used widely in schools in the United States into the nineteenth century. While public education has since moved away from many of the principles of classical education, such as the learning of Latin and a chronological approach to history, classical education has left a strong legacy. We still refer to early grades as grammar school, for example, and we have separate learning "stages" of middle school and high school.

In recent years, classical education has gained renewed popularity both in private schools and in homeschooling. Modern classical education is attractive to some homeschoolers for many reasons. It has its own built-in structure and plan of study, especially if parents use a guide such as *The Well-Trained Mind*, which offers a year-by-year study plan and extensive lists of resources. The three stages of learning—the trivium—does seem to follow many children's intellectual development. When used with flexibility and adaptation for the individual child, classical homeschooling is a good fit for many families and students, and it offers a reassuring educational plan and structure that sets the tone for lifelong learning.

Like homeschooling itself, classical education isn't for everyone. Some parents find the approach too rigid, or their children do not fit the developmental learning stages for which the trivium is designed. Other parents disagree with the Western-centered approach to history, languages, and literature and want to offer their children a more diverse early education, or they want a curriculum that focuses more on math and science. When classical learning does work, however, it can result in joyful, lifelong learners who are well prepared for further education and work. The authors of *The Well-Trained Mind* have also written a book for adults and older teens, *The Well-Educated Mind: A Guide to the Classical Education You Never Had,*[8] which shows lifelong learners how to give themselves an education in the Great Books and shows how to "jump into the great conversation" of classical learning.

Distance Education and E-learning

The World Wide Web and electronic learning have been responsible for broadening homeschool resources in ways that were unimaginable just a few years ago. Homeschoolers can take courses online, find information within minutes from anywhere in the world, even get a high school diploma through distance learning programs. In addition to distance learning programs that cater to homeschoolers, such as Clonlara School and Keystone National High School, several universities offer online high school courses and diplomas, such as the University of Nebraska High School, the University of Missouri–Columbia High School, and the University of Texas at Austin High School. Many of the programs give students a choice of either working from hard-copy textbooks and mailing assignments, or using online, electronic resources and submitting work through email.

The Internet is an invaluable resource for parents as well. Parents can use the Web to learn more about home education, to get curriculum ideas and download free lesson plans, and to connect with other homeschoolers through message boards and e-lists. A Yahoo search for homeschool interest groups yields over four thousand different groups![9] There is a specialized homeschool e-group for almost any need or interest, including homeschooling boys, thrifty homeschooling, homeschool resource swaps, homeschooling autistic children, and homeschooling for high school. Many parents and families have their own Web sites, where they record their family's homeschool journey and share their interests and resources.

Critics cite potential drawbacks of relying too heavily on electronic resources in homeschooling. They paint a picture of a lone and lonely child at a computer for several hours every day, communicating with the outside world through email and instant messages. I personally know of no families whose homeschooling looks like this. Judicious use of new technology and avoiding isolation are important issues for all children, not just homeschoolers. Children who attend school are certainly not immune to loneliness or to feeling like outsiders, nor do they always spend their free time in productive ways. Homeschool parents and school parents alike face unique challenges in the electronic age, as more and newer forms of entertainment vie for our children's attention.

THE PACE OF HOMESCHOOL DAYS

Pacing is an important aspect of any homeschooling approach. Allowing children to learn at a pace that is natural for them, whatever that pace might be, is one of homeschooling's greatest gifts, and is one of the important ways that homeschooling differs from classroom learning. Honoring children's natural learning pace gives them an edge that no amount of tutoring can buy: self-confidence and a positive self-concept.

Gaining a new perspective on how fast children should learn requires a lot of deschooling. Grace Llewellyn and Amy Silver, authors of *Guerrilla Learning: How to Give Your Children a Real Education With or Without School,* explain that wanting our children to keep pace with others stems from the best of intentions and is embedded in how we think:

Why is it so difficult for us to respect our children's timing? Because we want so much for our children and ourselves, and because we as adults are so often in a hurried panic about everything. Because rush is the air we breathe. Because we're fraught with worry that our kids will end up on the streets if they're the last in their class to learn their multiplication tables.[10]

What's the harm in expecting our children to keep up with what's expected of them? For children whose pace of learning is not in step with timetables of what all children should learn and when they should know it, an inflexible learning environment can feel torturous and can have a great impact on later education. Ned Johnson and Emily Warner Eskelsen describe how easily self-concepts are undermined when children think of themselves as flawed learners:

> Self-concepts about reading originate early. From elementary school, all children know which reading group they belong to, and therefore whether or not they are "good readers." Think of the rigid demarcations often seen in fourth grade classrooms— the A group reads thick story books, while the B group has thinner ones, and the C group leaves the classroom for remedial aid. The process continues throughout elementary school and into middle school, where often students are grouped into classes by level of reading ability. That's a lot of years of reinforcement, which can be difficult to overcome.[11]

I've met many homeschooled children who learned to read very early, and just as many whose reading didn't take off until age eight or nine or even later. What if those later readers had been placed in a C reading group? Seeing themselves as poor readers or nonreaders, would they have lived up (or down) to the expectation? Homeschooling allows these children to continue to love books through hearing them read aloud until the time when they are ready to read for themselves. Of course, some homeschooled children benefit from specific reading strategies, and this may be a challenge for some parents. Parents can pay close attention to how their children are reacting to learning to read, especially if the child shows frustration, and, if necessary, do research to learn ways to make the reading process easier.

Many children in classrooms are either constantly trying to catch up or marking time until school is out, when they can do some real learning. School parents of late bloomers and early bloomers—as Llewellyn and Silver call them—spend much of their energies and time on the issue of pacing. To ensure that their children are neither behind nor bored, they meet with teachers, oversee homework, seek remediation, hire tutors, and work to get their children in honors programs. They also worry—a lot.

Our concerns as parents are often based more on unfounded fears of the future than real, present problems. In the back of our minds, we worry that a child who isn't reading in first grade won't get into a good college, or that a gifted freshman who isn't placed in honors geometry has ruined any chance of fulfilling her potential. But if we look at such children without the distorting lenses of future worries or comparisons with other children, we can see that there is often no problem at all. The six-year-old will probably learn to read next year if not sooner, and the high school freshman might be just fine in non-honors math, but only if we can step back and see the situation clearly.

Homeschooling offers a different and refreshing perspective on pacing, if we are willing to let go of many of our assumptions about children and learning. Let's look first at children who are late bloomers. Keeping the botanical metaphor, these young learners will flower later than other children. Like chrysanthemums, they might need a long growing time and even some pruning along the way to allow for their glorious fall displays. Late bloomers might learn to read later, grasp math facts later, or develop social skills later than age peers. It's just how they grow. Early bloomers also grow differently. They come up early in the spring, pushing their way through April snow like hyacinths. Their timing is internal, and if they are pruned when young they may not bloom at all. If allowed to grow according to their needs, however, they are hardier than they look and provide glorious color long before other shoots have poked through the earth. It's just how they grow.

And here's where the metaphor ceases to be useful, because children are not plants.

They are not trimmed back once they bloom nor do they need to be divided or replaced every few years (although we do assume a certain amount of dormancy in the summer months). Children who are late bloomers are right on time for their own development, and they show beautiful color long before they are said to blossom. Children who are early bloomers don't use up all their potential at the beginning, and, unlike perennials, they can continue to flower nonstop throughout many seasons to come. These children are neither early nor late in comparison with where they are supposed to be, but in the classroom, their growth and development will always be compared with an ideal average.

ECLECTIC HOMESCHOOLING: DOING WHATEVER WORKS

Figuring out how to homeschool is much like learning how to be a parent. Being a good parent depends in large part on our individual skills and beliefs, our children's personalities and needs, and our unique family structure and circumstances. Most people would agree that there is no one right way to parent, because no two families are exactly alike. Similarly, there is certainly no one right way to homeschool. It is important for parents to give themselves permission to use whatever works for their child, rather than follow the latest trend or feel pressure to do what works for someone else. Like parenting, homeschooling is a learn-as-you-go enterprise, constantly adjusting itself to the growing child and changing family. Like good parents, effective homeschoolers learn to find a comfortable balance between "book learning" and gut instinct.

In the end, many homeschoolers find that labels for different homeschooling styles can get in the way of learning and education. Like many other homeschooling families, we have found that, over time, labels cease to matter and we think of ourselves simply as learning at home, using whatever works from day to day, year to year, subject to subject. This approach is sometimes described as eclectic or relaxed homeschooling, and is probably how most homeschoolers would describe how their children learn. Of course, it's not a single approach at all, but rather attentive and conscientious parenting extended to a child's home education.

> *Our approach to homeschooling is what I'd describe as "relaxed"; it's a style that perhaps falls midway between "unschooling" and "school-at-home." I see myself in a facilitator role rather than that of a teacher. I try to gather the tools needed for my kids to succeed in their quests, prepare them for life, recognize needs, encourage them to explore beyond their established comfort zones and serve as a sounding board. As our children investigated areas beyond my expertise (and even interest), I found myself being a student or co-learner alongside my children. This, I learned, is my favorite place to be as a homeschooling mom.*　　—Kathleen

Eclectic or relaxed homeschooling, perhaps more than any other approach to learning at home, will be very different from family to family, and even among siblings. This is true individualized learning. Becca, who has homeschooled for eight years, offers one example of how eclectic homeschooling works:

My approach has morphed many times but it has always fallen under the category of what I term "eclectic" homeschooling. Basically, I use whatever works! This would include conversations, books, working with tutors, local classes, curriculum, videos and DVDs, museums, online classes, cooperative groups, and more. I also try to change whatever isn't working. I have periods of little or no structure and times when my sons are following some sort of schedule. I've always followed my children's interests and tried to supply them with resources to facilitate their passions and interests. I also encourage areas of study that my husband and I deem important such as Bible study, self help skills, the basic three Rs, proficiency in a foreign language, and other things, while being sensitive and flexible according to each child's unique learning preferences and abilities.

Also, if there are areas that I think are important but my children don't have a particular interest in (such as history), I find ways to make it interesting and palatable to them and allow them to dabble without feeling forced to learn something they'll forget anyway. I also believe in "shared learning," that is, sharing what we learn with each other and enriching our lives in that way. Some of our best learning moments come from lively discussions or debates about something we've read or heard.

I don't follow a grade-based curriculum for the simple reason that neither of my older sons is working straight across the board at his grade level. Each boy knows what grade he would be in if he were in school, but at this point we don't really keep track of grade levels of materials for my oldest son and only in math for my middle son (because it's labeled on one of the books he uses). For record keeping and competition purposes (my oldest is involved in a number of activities that require me to put his grade level), I do have my children at their same age or grade level as their peers.

For my middle son, age nine, I've never used any evaluation until about a month ago. I've never kept grades for him. I gave him a standardized test at home for the first time simply so he could have the experience of taking this type of test. For my oldest, I've used lots of evaluations but not for every subject. For his work for outside sources such as his online college class and his high school literature classes, he automatically gets evaluated just like every other student in these classes. He's also taken the SAT I, and I've done at home standardized testing with him every other year.

Becca's approach to home learning is clearly driven by her children's and family's needs. She also shows that homeschooling is almost never the same from year to year for a specific child. One advantage that home education offers is that parents can switch gears whenever it is necessary. If a certain approach is not working well, there is no need to wait until a new academic year or term to make necessary changes. Free from the bureaucracy of government and other conventional forms of education, children are able to receive an education tailored to their current needs.

Orlando and Tim tell how approaches to learning at home can change, and how families can feel free to adopt and adapt from different styles to find what works, regardless of what it is called:

We followed a fairly structured homeschool in the earlier days, what I would call "structured homeschooling." We did lots of worksheets, book reports, field trips, and followed a fairly standard curriculum. In high school, we transitioned to a more "structured unschooling" approach, in that I took what I wanted but followed a fairly structured curriculum based on my desires in education. My mom has always been a part of my learning, although in high school her involvement dropped off sharply, as she allowed me to explore learning in my own ways and teach myself a good deal of what I have learned.—Orlando

[Our homeschool approach was] partially college prep, partially unschooled. Towards the end it was more college prep than some of the people we knew. We would finish textbooks when we finished them, instead of sticking to a particular schedule. Of course, as I got older, I

was more inclined to stick to the same sort of schedule as my peers.
But we used those textbooks to their full potential.   —Tim

SUMMARY: LIMITLESS OPTIONS FOR LEARNING

Homeschooling offers limitless possibilities for how to learn. Combine this with family differences in size, geographical location, talents, goals, income, values, and special needs, and it's easy to see that there is no one right way to learn at home. In classrooms, all the children often do have to follow the same approach for the sake of efficiency. It would be impossible to tailor each subject to each student's learning style, pace of learning, and interest. In homeschools, individualized learning is not only possible, but can happen every day, for every subject.

Does it matter which approach is used, or if any single approach is followed consistently? Not really. Yes, it pays for parents, caregivers, and children to consider options available to them that can make their home education easier, based on their needs as a family, state requirements, learning styles for both adults and children, time, money, and other resources. If a specific approach to homeschooling works, it makes sense to use it! On the other hand, what works for a family can change from year to year and child to child, so flexibility is key.

The truth is that it's okay to use no specific approach at all, other than following your child's lead and striving to provide the best education possible. Some parents, such as Becca, take great joy in creating their own homeschool approach, dealing with individual issues and learning needs as they arise: "I love to research materials, curriculum, and classes and... to modify, add to, change, adjust, and create what we use for learning. Having been a deaf education teacher, I have been accustomed to innovativeness and individuality in learning; it just makes sense to me." One of the secret joys of homeschooling for parents is the creativity they can use as they educate their children.

Once again, the role of good parenting seems to take center stage, as Amanda Petrie, home-education researcher at the University of London, concludes: "The effectiveness of home education would seem to depend, not on the methods which the parents employ, but rather on the level of commitment of the parents to their children's educational needs."[12] Almost any homeschool approach can work if it is a good fit for the parent and child and if it is followed with flexibility and in service of the pleasure of becoming an educated person.

DOING YOUR HOMEWORK:
SELECTED READING

Becca, who advises other families who are thinking of homeschooling, suggests that parents research all sorts of homeschooling approaches to have the best chance of meeting the needs of their children: "Even if a parent is inclined to be very structured, I would still encourage them to read a variety of authors from the whole homeschooling spectrum, from unschooling to school at home." The following resources offer a broad view of what homeschooling can look like and ideas for how children can become educated at home.

Freedom Challenge: African American Homeschoolers (Lowry House, 1996), *The Teenage Liberation Handbook: How to Quit School and Get a Real Education,* revised and expanded edition (Lowry House, 1998), and *Real Lives: Eleven Teenagers Who Don't Go to School* (Lowry House, 1993), all by Grace Llewellyn. Llewellyn is one of the most down-to-earth, friendly voices in the homeschool world. Her emphasis is on self-directed learning, and her books would be helpful for any family who wants a better understanding of homeschooling and whether it can prepare teens for life after high school.

Real-Life Homeschooling: The Stories of 21 Families Who Make It Work, by Rhonda Barfield (Fireside, 2002). For anyone who wants a peek into the lives of homeschoolers, Barfield takes you into their living rooms and communities to show what day-to-day home education looks like. The families who are profiled represent a broad spectrum of political views, religious beliefs, educational approaches, and ways of life.

The Unschooling Handbook: How to Use the Whole World as Your Child's Classroom, by Mary Giffith (Three Rivers Press, 1998). Griffith writes, "Unschooling, to me, means learning what one wants, when one wants, where one wants, for one's own reasons." In this easy-to-read introduction to unschooling, she shows exactly what a self-directed education looks like, with many examples and an unbridled enthusiasm for learning outside of schools.

Home Learning Year by Year: How to Design a Homeschool Curriculum from Preschool through High School (Three Rivers Press, 2000) and *The Complete Home Learning Source Book: The Essential Resource Guide for Homeschoolers, Parents, and Educators Covering Every Subject from Arithmetic to Zoology* (Three Rivers Press, 1998), both by Rebecca Rupp. For parents who want a more structured approach to homeschooling, *Home Learning Year by Year* offers an extensive overview of what is normally expected to be taught in each grade and ideas for lessons and resources. Rupp's *Complete Home Learning Source Book,* while a few years old, is nonetheless a valuable glimpse into the world of homeschooling resources (readers can check if an updated edition is available).

Homeschooling in Wisconsin: At Home with Learning, by the Wisconsin Parents Association (Wisconsin Parents Association, 2006). Parents who want an inclusive view of and resource for homeschooling that is unbiased toward individual perspectives can read this book. More than just a guide for Wisconsin homeschoolers, the handbook offers detailed suggestions for getting started, making use of free and inexpensive resources, pursuing home education through high school, and many other important issues. Susan, veteran homeschool parent of three children and a WPA regional coordinator, recommends this book to anyone new to homeschooling, regardless of where they live:

> At Home With Learning *is a wealth of information and wisdom that I use all the time myself. Without being authoritarian, it suggests ways to approach family life and homeschooling that empower each person or family to make the choices that are right for them. Homeschooling does not look the same for every family. Some choose to purchase a curriculum and follow it exactly while others look for a variety of interesting books and life learning experiences to suit their needs. Both of these approaches (and everything in between) are fine and that, in my opinion, is why homeschooling works so well: we aren't locked into any one approach.*

5

HOMESCHOOLING FOR SPECIAL NEEDS

It is a daily struggle for the average child to get an education when classrooms are overcrowded, learning materials are outdated, and overworked teachers have been reduced to college-educated childcare providers. For those children who are in any way different from the norm, it is virtually impossible to learn in such a chaotic and unforgiving environment.

— *Lenore Hayes,* Homeschooling the Child with ADD (or Other Special Needs)

DIFFERENCE OR DISORDER?

When are a child's learning or social differences far enough from the norm to be termed special? At what point is a child who is easily distracted termed to have an attention disorder? What is the line between being very sensitive to noise and at risk for autism? When does a late reader become a potential dyslexic?

Some children have differences that are striking. From an early age, they react to ordinary situations in ways that are extreme when compared with other children their age. They struggle with transitions. They learn at a much slower or faster rate than the rest of the class. They are socially and painfully out of step with age peers. Nearly all of their energies in the classroom are spent dealing with internal and external adjustments rather than the joy of learning. Some special needs are ADD/ADHD, autism, Asperger's syndrome, sensory integration dysfunction, and learning disabilities such as dyslexia. Highly and profoundly gifted children also can require accommodations quite different from those provided in regular classrooms, or they may be "twice exceptional": gifted and with learning disabilities.

Parents of children with special needs often turn to homeschooling not as a first choice but, in the words of Lenore Hayes, "by default." Hayes describes in *Homeschooling the Child with ADD* how her son had been in three part-time preschools by age four because he did not conform to the expectation of switching activities every fifteen minutes. He would just be getting interested in whatever he was doing when he'd be asked to drop it and go to the next task. His parents were advised that at his age he should be able to handle the transitions, and that he probably had attention deficit hyperactivity disorder (ADHD) or autism. When it was time for kindergarten, they felt "stuck," with no option left to them but homeschooling. To their surprise, homeschooling finally gave their son the educational environment he needed:

> That once elusive diagnosis for our son no longer mattered because we had the freedom to work with his differences and make adjustments as needed. There was no need to publicly place his weaknesses under the microscope of school administrators. We never had to jump through bureaucratic hoops to obtain what we needed for our son's education. As a one-income family, we were able to create a vibrant learning environment using library books, art supplies, field trips, and other low-cost resources. Homeschooling has been very beneficial for my square-peg-in-round-hole son.[1]

For children who feel like square pegs, homeschooling can be more than a good fit. It can mean the difference between a miserable child and a happy one. Hayes's story is familiar to many parents for whom homeschooling is a last resort. For some of these families, homeschooling offers a temporary break from inappropriate classroom environments, giving the parents time to research other options and the children time to recover confidence in their abilities. For other families, homeschooling continues to be the best long-term plan.

BUILDING CONFIDENCE: USE LABELS WISELY

How do your children feel about how they learn? Do they attach any labels to themselves? Do they think of themselves as smart? Young scientists? Lovers of language? Seekers of knowledge? Or do they say things like, "I don't test well" or "No one in my family is good in math" or "I have trouble paying attention"?

All parents of children with extreme learning needs—whether they send their children to school or homeschool—should think about the role of labels in their children's self-concept. Homeschooled parents are in a better position to have control over the use of labels. The authors of *The Mislabeled Child* warn that "a label should never be used as a shorthand for a child's whole existence. Statements like 'He's ADHD' or 'She's Asperger's' leave far too much unsaid to convey a complete or accurate picture of a child."[2] A *Wall Street Journal* article asks if "the stigma of being labeled with ADHD could lead some kids to lose confidence, and dream smaller dreams."[3] In school, children may have no choice but to live with labels or be singled out for their differences, such as ADHD children being required to move their desks to the front of the room, near the teacher. At home, parents can choose whether and to what extent to emphasize and label individual differences or to accept them as part of being human.

The impact of labels is felt both by children who struggle and by those who zoom ahead. I once asked a panel of homeschooled gifted teens how they felt about the word "gifted," and every one of them said they didn't like it. Being gifted is just a part of who they are—like having green eyes or freckles—and they don't want to be singled out or praised for something over which they have no control and that they didn't earn. They also said that they know they can't control what other people think of giftedness, and they prefer not to be prejudged. These children all knew they were gifted. They were well aware of their differences, but they didn't think the word "gifted" helped others to understand them any better. In fact, it sometimes gets in the way.

Avoiding unnecessary labels does not mean that parents who homeschool pretend their children are perfect or that they ignore problems. They simply know that confidence breeds success, and when children feel confident about their learning—regardless of how far they are from the norm—they are more relaxed, and they learn more easily.

HOMESCHOOLING FOR SENSORY CONTROL

Many children with Asperger's Syndrome, other autistic-spectrum disorders, sensory-processing disorders, and high levels of giftedness and sensitivity learn effectively and more enjoyably at home because parents can control the sensory environment. In *Choosing Home: Deciding to Homeschool with Asperger's Syndrome,* Martha Kennedy Hartnett describes how parents whose children have Asperger's Syndrome often have to reteach what their children were taught at school, because the normal environment of the classroom prevented them from learning.[4]

As many as 10 percent of children have difficulty processing sensory input to an extent that interferes with classroom work or behavior.[5] These children show high sensitivity to lights, sounds, smells, touch, movement, and other stimuli. Even the usual hustle and bustle of a classroom of two dozen young children may provide enough distraction to impede learning and enjoyment. The usual noise and smells of a crowded cafeteria or the frenzy of a recess playground can cause these children to shut down in self-defense or lash out in frustration.

People who are not highly sensitive to their environment may assume that these children can just get used to it, learn to have a thicker skin. However, it's not a matter of mere self-control. Lisa Pyles, in *Hitchhiking Through Asperger Syndrome,* offers this exercise to show what daily life in school can be like for a child with extreme sensitivity to stimuli:

> Imagine having the "volume turned up" on every sense. Imagine a day when you have premenstrual tension (for male readers, imagine early flu symptoms). Halogen lights are in your eyes. Someone is squeaking a balloon and it's grating on your last nerve. Meanwhile you're expected to balance a checkbook.
>
> Now think about sitting in the middle of a parking garage. Cars rumble past inches from your feet. You have mild sunburn all over. Exhaust fumes are turning your stomach. A bird swoops and chirps too close to your ear. No place feels sturdy enough to write on but you are expected to compose a business letter.[6]

Children who have problems processing normal sensory stimuli experience this kind of struggle much of the time, and as a result they may go into fight-or-flight mode in environments the majority of people consider inoffensive. Because such children are unable to articulate what they are feeling (they don't understand it themselves), their extreme discomfort can easily be mistaken for rebelliousness, developmental delays, or antisocial behavior. This calls attention to the problem, but doesn't always lead to the help the child needs. Other students don't have as severe difficulties and seem to cope—they do most of their work and don't have behavior problems—but they are always a little on edge, and all of their internal resources go toward just getting through the day.

Many highly gifted children in regular classrooms also struggle with sensory input, and they are unable to learn at a level that brings them joy and challenge, not because they don't want to, but because their emotional and physical sensitivities get in the way. They simply notice much more than everyone else. The authors of *A Parent's Guide to Gifted Children* write that gifted children commonly display "extreme sensitivity to various kinds of sensory stimuli."[7] Adults who do not understand that this is often part and parcel of being gifted can misdiagnose the child or even shame the child for being "too" sensitive.

Young sensitive boys, in particular, who might be expected not to let things bother them and to be emotionally tough, can benefit from the safe and affirming environment of home. Orlando, now eighteen, says that he was bothered by the rowdiness of the classroom. His parents decided to homeschool him when he was six:

> *My parents knew me as a sensitive person who hated certain aspects of the school environment. They thought that putting me in a school setting would not be beneficial to my mental health, and have used the phrase, "You would have been eaten alive" to describe what might have happened.*

Homeschooling allows boys like Orlando to become comfortable with themselves and their sensitive natures in a culture that assumes boys should be competitive, enjoy rough play, and be tough enough to withstand teasing.

Whether their children are extremely sensitive or have a diagnosed sensory input difficulty, parents can think about whether a calm homeschooling environment might help them to learn more efficiently and with more enjoyment. Because school-age children are often in stimulating environments for most of their waking hours, parents might need to observe their children during vacation or at times when they are in quiet and peaceful surroundings. Are they noticeably more focused? Happier? Better able to learn? What might their education be like if they could learn this way most if not all of the time?

SPECIAL SERVICES FOR SPECIAL NEEDS

For children whose needs are extreme, especially in geographical areas with limited schools or services, homeschooling may be one of few options. Parents who are thinking of homeschooling a child who requires specialized care and education can learn about how other parents have managed their homeschooling and what resources are available. Parents who have children with diagnosed special needs will want to know what their state law says about their children's rights, and if any state and school services are available to them as homeschoolers.

Parents can be pressured by school officials or well-meaning friends and relatives to enroll a child with special needs in school. Some children require therapeutic services hard to obtain or financially prohibitive outside of school, and the intensity of homeschooling a child with extreme needs can leave parents drained. However, parents should not assume that homeschooling is not an option for them. The most important thing is to meet the child's needs as fully as possible. Sometimes this requires a combination of homeschooling and other services, either public or private. And sometimes homeschooling is not the best option at a given time. The Internet resources at the end of this chapter offer more information on homeschooling for specific special needs.

MORE-ORDINARY NEEDS

While some children's special needs are easily diagnosed, other children's learning requirements are less clear. They seem to learn differently from most other children, but is it different enough to warrant special attention? Parents may wonder if seeking individual accommodations is worth the inevitable testing and labels. And what about the children whose abilities and learning styles place them in the middle of the bell curve? Do they also deserve an individualized education? Or will they do just fine if left alone?

One advantage of homeschooling is that every child's needs can receive individual attention, regardless of labels or diagnoses. This sounds like a platitude, but it is a reality. To provide children with special needs the education they deserve, homeschooling parents often do extensive research, becoming experts in what their children need to thrive. They set up support networks with other families and make accommodations in their homes and lives that would be impossible in the classroom. One parent I know set up a sensory integration area in her basement, complete with an indoor swing, mini trampoline, and large rubber balls. Her young daughter was able to get the physical therapy she needed—all in the name of fun—while learning at home.

For children whose needs are less intense, homeschooling offers many ways for parents to individualize learning and meet specific needs: by paying attention to how children learn best and discovering their learning styles, by letting children learn at their own pace, by controlling the environment to reduce distractions, by taking care of physical needs, and by building self-confidence.

LEARNING STYLES

How do you prefer to follow directions? Using an Internet site such as Mapquest, do you use the map or the written directions? Can you follow directions if someone reads them aloud to you, or do you need to read them yourself? For some people, it doesn't matter much one way or another, but for others, the difference between learning with pictures and learning with words could be the difference between getting to the destination on time or getting hopelessly lost.

There are many different ways to understand learning styles. One of the most popular models distinguishes between learning by seeing (visual), learning by hearing (auditory), and learning by doing (kinesthetic). Think of how children play with small plastic building blocks. They might prefer to follow picture-based instruction, especially if they can see a picture of the completed project (visual). Others do better if someone tells them how to put the pieces together (auditory). Or they may prefer to dive in and figure it out as they go, learning by doing (kinesthetic). Some children strongly lean toward or have particular difficulty with one of the three modes of learning, and others can shift easily between all three.

In a homeschooling environment, parents can use the idea of learning styles to offer children a range of ways to approach a single topic. For example, children can learn history through reading, through conversation and listening to documentaries, and by visiting living history museums and doing history-based crafts and activities. By sampling all these ways of learning, children use their strengths to their best advantage and practice learning in less preferred ways. Parents who watch how their children respond to each way of learning are in a good position to continue to offer learning approaches that meet their needs.

Howard Gardner's multiple intelligences approach is another way that parents can tailor education to a child's preferred ways to learn. Gardner presents eight ways that we can use and show our intelligence: through words (linguistic intelligence), through numbers and reasoning (logical-mathematical intelligence), through music (musical intelligence), through pictures (spatial intelligence), through our bodies and movement (bodily-kinesthetic intelligence), through our relationships with other people (interpersonal intelligence), through knowledge of and a relationship with ourselves (intrapersonal intelligence), and through nature (naturalist intelligence). Because schools traditionally value and reward linguistic and logical-mathematical intelligences the most, children who show their intelligence in other ways can find themselves not tapping into their potential or even misdiagnosed with learning disabilities.

All of these models offer explanations and descriptions of how we prefer to take in and express knowledge. Some children may not have a strong preference, or may be able to tolerate environments not suited to their preferred learning styles. Others may become frustrated to the point of misbehavior or apathy, especially if they do not have the chance to learn through the style that serves them best. Children with short auditory attention spans often struggle in classrooms in which children are expected to listen to the teacher for extended periods of time. Likewise, young children who prefer to learn by listening rather than by doing may balk at the physical games of preschool and would enjoy, instead, being read to or engaging in conversation with adults.

There is some controversy about the validity of the science of learning-style theories. Some doubt the familiar left-brain–right-brain model, which holds that left-brained thinkers are more logical, sequential, and auditory (think of accountants, proofreaders, and lawyers) while right-brained thinkers are intuitive, random, and visual (think of architects, artists, and actors). Critics say that this model is too simplistic, and that barring rare organic defects or traumatic injury, we all use both halves of our brains all the time. The main idea of the theory, however, remains extremely useful, especially for anyone who works with children. We all know families in which it's easy to see who has left-brain strengths and who has right-brain strengths. Anything that helps us understand and respect those differences rather than assume that everyone should be the same is helpful.

While researchers debate the finer points of learning-style theories, parents and teachers find that paying attention to how children best absorb and process information can make a big difference in the quality of and attitude toward learning. For example, I know many children who can listen more attentively if they also have a way to keep their hands busy, with small toys or play dough. Color-coded spelling words can help visual children to retain what they learn. Learning math with manipulatives (objects such as dice, wooden rods, and playing cards) is more effective in many cases than using a textbook or workbook. And some slow readers who are strong auditory learners can use books on tape to hear books they easily comprehend but that are otherwise beyond their ability to read.

Becca, who homeschools three boys, calls these kinds of accommodations "learning helps":

Learning styles and learning differences have played a huge role in our homeschooling. From the time my oldest was four or five, I began to research and study how he learned. I read several dozen books on "learning helps" as I call them, and have even given talks at homeschool conferences on what I've learned. I still continue to thoroughly enjoy learning theory research, and I try to incorporate what I know into each child's education.

What's fun is that each of my sons is really different in how he learns. The learning style or strength in one child allows us to expand our territory and strengthen weaknesses. For example, my middle son is extremely auditory and has a long attention span (fairly opposite of my oldest son). He has listened to long novels on tape since he was five. My oldest has increased his auditory abilities as well as his attention span by also listening to these books on tape.

When parents begin to understand how their children learn best, they can unlock learning potential and bring enjoyment to the learning process. As Becca shows, having children in one family with different preferred learning styles is not a problem. By approaching subjects in a variety of ways, children not only strengthen what they already do well, but they also get valuable practice in ways of learning they may not choose on their own.

Attending to children's learning styles has benefits for children of all ability levels. Highly gifted children have been found to prefer to learn through discovery and problem solving, self-direction, and variety.[8] Homeschooling makes room for these preferences by allowing children the time to discover truths and solutions for themselves, giving them the freedom to choose what to learn and how to learn (at least some of the time), and allowing them the space to move to a different topic or subject when they need a change of scenery.

Parents who homeschool also learn something about their own learning styles as they work with their children. Dad might see in his daughter a gift for visual learning that his own parents had discouraged, but that he now can rediscover. Mom can learn that, just like her son, she learns best by doing (that's why she always skips the directions when cooking!). Parents also sometimes realize that their learning styles are very different from those of their children. Such a mismatch of learning styles—for example, a child who is strongly visual-spatial in an auditory-sequential family—requires some flexibility and adjustment to make sure that adults' preferences aren't automatically thrust upon children. If a child's learning style is markedly different from the rest of the family members, he or she might benefit from an older child or adult mentor who shares that style and can show the child how to use and appreciate it.

When we give children the chance to learn in ways that are best for them, we boost their confidence as well as their achievement. Rather than focus on labels that carry negative connotations and imply that something is wrong with the child, homeschool parents can choose to speak of learning strengths. Not only does this make children feel good about what they do well, but they are more willing to practice less preferred ways of learning, knowing this is not an effort to "fix" them.

THE PHYSICAL SIDE OF LEARNING

Common sense tells us that children will learn best when they are well nourished, getting enough sleep, and engaging in regular physical activity. These needs are basic to all children, yet in 2007, the *American Journal of Preventive Medicine* reported that 55 percent of eleven-to-fifteen-year-olds did not meet physical activity guidelines. Twelve percent got enough fruits and vegetables, and only 32 percent met recommendations for fat intake.[9] Several studies show that most children do not get the sleep they need, especially in adolescence, when, paradoxically, sleep needs increase at a time when teenagers like to stay up late and need to wake up early for school.

One reason that Elizabeth and Dan Hamilton decided to homeschool their five-year-old daughter was to control her allergies and asthma. School regulations did not allow her to carry her asthma medication with her, and her parents were concerned that she might not get to the school nurse soon enough if she had an asthma attack.[10] Families with children who have severe allergies, such as peanut allergies, or other health problems find that homeschooling can be a healthier and safer option than the classroom.

Just as sleep and diet requirements differ from person to person, some children need more physical activity than others. Especially for many young boys, homeschooling allows for the movement they seem to crave without the need to stay in a seat most of the day, stand in line quietly for reasons they don't understand, or wait their turn with twenty other children for a drink of water. Becca says of her son, the thirteen-year-old who is now taking online college classes, "My son could never sit still in the classroom. He would have been labeled ADD right from the start."

HOMESCHOOLING GIFTED CHILDREN

No other educational option accommodates the wide range of abilities of gifted children as well as homeschooling. Gifted homeschoolers can move ahead without skipping a grade. They can learn math at high school level while enjoying sixth-grade books. Or they can take their time through high school and study topics of interest in depth rather than going to college early. A few profoundly gifted children homeschool because they are ready for college-level work before they reach middle school. In all cases, these children fit perfectly in no single grade, and homeschooling doesn't force them to try.

For children who are precocious learners, homeschooling offers unique benefits and a few challenges:

> *A challenge has been to find appropriate materials and learning situations for my oldest son. When he was younger, he was an advanced reader but highly sensitive and I found it difficult to find material at his reading level that was socially and emotionally appropriate. Because he is so advanced in certain subjects such as math, I've had to do hours of research to help find the best learning situations for him. Even homeschoolers face age-based prejudices. . . . I plan on homeschooling my children through high school (and beyond).*   —Becca

Parents like Becca learn to make homeschooling fit the needs of their gifted children rather than forcing their children to fit into a school model. Many people assume that gifted children are advanced evenly in all areas, so there should be no problem with simply letting them skip a grade, but this is not the case. They might be years ahead of other children in one or two subjects, but at grade

level or even below grade level in others. Some are advanced in almost every subject academically yet their sensitivities prevent them
from socializing with older children. They have a hard time fitting
into any one group comfortably.

Parents do not have to know more than their gifted children in
every subject area to homeschool successfully. Parents need to be experts only about their own children's learning styles, personalities,
and educational needs. When a child surpasses the parent's knowledge base in or ability to teach a particular subject area, his or her
needs can be met through distance-learning courses, talent search
programs, mentors, community classes, homeschool co-ops, or part-
time enrollment in public school, private school, or college. More-
over, many gifted children can learn to teach themselves what they
want and need to know.

Parents who homeschool highly gifted children and other children with special needs learn to do a lot of research to find the resources and assistance their children deserve. They know they
usually can't do it all themselves, and they become experts at discovering solutions that others have overlooked. They also put a high
priority on their own well-being, know how to ask for help, and find
ways to give themselves a break from the intensity of working with a
child whose needs are extreme.

THE HOMESCHOOLING DECISION:
MEETING THE SPECIAL NEEDS CHALLENGE

If you are the parent of a child with special needs, only you can
know if homeschooling is a good option for your family. As many of
the examples in this chapter show, homeschooling could unlock
your child's potential, simplify your life, make day-to-day routines
easier for everyone, and allow your child to learn with more joy. On
the other hand, being with a challenging child all day, every day can

be a challenge in itself. Think about the following questions as you decide what is best for you and your child:

- Is my child old enough to have an opinion about home-schooling? If so, what does he or she think?
- Do I want to homeschool? Does my spouse want to home-school?
- What do my state's laws say about homeschooling children with special needs?
- Will I be able to meet my child's needs? For needs that I can't meet, can I find someone who can help?
- Do I know how my child prefers to learn? Would homeschool-ing allow him to use preferred learning styles more often?
- Would homeschooling help my child to learn according to her natural rhythm and pace?
- Would I be able to provide a calm learning environment for my child?
- Would homeschooling have an effect on my child's sleep habits and overall health?
- How might homeschooling affect my child's self-concept?
- How do I feel about labels for my children?
- What is the best possible outcome of our homeschooling for one year?
- What could go wrong if we homeschool for one year?

DOING YOUR HOMEWORK: SELECTED READING

Selected Web Sites

Home School Association of California: Homeschooling with
   Special Challenges
http://www.hsc.org/chaos/specialchallenges/

Illinois H.O.U.S.E.: Homeschooling the Special Needs Child
http://www.illinoishouse.org/a13.htm

Bay Shore Educational: Homeschooling the Learning Disabled and
    Other Special Needs Children
http://www.bayshoreeducational.com/special.html

NATHHAN National Challenged Homeschoolers Associated
    Network: Christian Families Homeschooling Special Needs
    Children
http://www.nathhan.com/

A to Z Home's Cool Homeschooling: Special Needs
http://homeschooling.gomilpitas.com/weblinks/specialneeds.htm

Homeschooling Kids with Disabilities (Yahoo group)
http://groups.yahoo.com/group/hkwd/

Homeschooling Your Deaf or Hard of Hearing Child
http://www.deafhomeschool.com/

Homeschooling Children with Down Syndrome
http://www.altonweb.com/cs/downsyndrome/index.htm?page=hom
    eschooling.html

Selected Books

*Homeschooling the Child with ADD (or Other Special Needs): Your
    Complete Guide to Successfully Homeschooling the Child with
    Learning Differences,* by Lenore Colacion Hayes (Prima Publish-
    ing, 2002)

Homeschooling the Challenging Child: A Practical Guide, by Christine M. Fields (Broadman & Holman Publishers, 2005)

Choosing Home: Deciding to Homeschool with Asperger's Syndrome, by Martha Kennedy Hartnett and Stephen M. Shore (Jessica Kingsley Publishers, 2003)

Home Educating Our Autistic Spectrum Children: Paths are Made by Walking, by Terri Dowty and Kitt Cowlishaw (Jessica Kingsley Publishers, 2001)

Homeschooling the Child with Asperger Syndrome: Real Help for Parents Anywhere and on Any Budget, by Lise Pyles (Jessica Kingsley Publishers, 2004)

The Out-of-Sync Child: Recognizing and Coping with Sensory Integration Dysfunction, by Carol Stock Kranowitz (Perigree, 1998)

6

TEENS AT HOME

Homeschooling for High School

Teens who can learn on their own attain educational nirvana—
they direct their own studies and teach themselves.
— *Cafi Cohen,* Homeschooling: The Teen Years

*Most lingering questions and worries naturally fell away as our older
children walked through the doors into their teen years, where I
found them delightful to be with. They emerged as positive, fun, well-
grounded, balanced individuals. This seemed a good sign to me.
When they were accepted into colleges, did well academically, socially,
and emotionally, again, we sighed. The clincher now is when I learn
from them that they are making it a priority to homeschool their own
(future) children.*
— *Kathleen*

GRADUATION DAY

It's high school graduation day. Parents and grandparents and
friends sit in auditorium chairs. Video cameras are set up to capture
the ceremony. The graduates take their turns walking to the
podium, accepting a certificate, saying a few words about their

appreciation for their education and anticipation of their future. Before attending college, one of them plans to take a year off, during which she will enroll in a few part-time classes for one semester and travel with her mother in the spring. Another talks excitedly of all she learned in her high school years and how she looks forward to pursuing an engineering degree in the fall. A third speaks of his interest in economics and other topics and their usefulness as metaphors for his education. Every graduate has a story to tell, and each has a chance to speak. Sometimes parents or a sibling also say a few words. There is no valedictorian. No honor roll. No one remembers high school as an unpleasant experience. All express gratitude for their education thus far as they look ahead to the commencement of their continued education—in school or out.

This is a graduation ceremony at a state home-education conference. The graduates have come to this milestone on many different roads. Some homeschooled for all of their elementary and high school education. Others went to school for a few years. One even had attended high school for her junior year after several years of homeschooling, then decided to finish high school at home for her senior year. Some pursued rigorous college preparatory study. Some followed their interests rather than a formal curriculum. All received a high school education unique to their abilities, personalities, and interests. At this ceremony, the students' accomplishments are validated not by a school official or a governing body, but by themselves, their families, and their friends.

HIGH SCHOOL AT HOME

When our son was about thirteen years old, friends and family members began to ask, "He *will* go to high school, right?" For awhile, we weren't sure ourselves. As I discussed in chapter one, we thought the matter over and our son visited the local public high school. In the end, he decided to continue his education at home. Homeschooling for high school is both different from and easier than I had expected, as it provides a high school education that he could get nowhere else, and nurtures our family relationships at a time when many parents and children struggle to stay connected.

"What about high school?" is perhaps the second most commonly asked homeschooling question after "What about socialization?" People have many reasons for assuming that homeschoolers will switch to classroom education for high school. Here are some common assumptions:

- Homeschoolers have trouble getting into college.
- Everyone in this society needs a high school diploma to be successful.
- Parents will have reached a ceiling on what they are able to teach, especially in math and science.
- Homeschooling can't offer appropriate academic challenge.
- Teenagers are too rebellious to cooperate with their parents for home education.
- It's too hard for homeschooled teens to find friends.
- Homeschooled adolescents miss out on sports and other group activities.
- High school isn't a good time to start homeschooling.
- And, finally, homeschoolers will regret not having gone to prom.

People naturally make such assumptions based on their own and their children's experiences of public and private schooling, and without having an idea of what high school at home looks like on a daily basis. The reality is that homeschooling through the high school years is not only practical, but also is easier than ever before and offers advantages unavailable in conventional secondary schools. To help you to know if homeschooling is the best option for your teen, let's look at some realities of bringing high school home.

DIPLOMAS, TRANSCRIPTS, AND OTHER PIECES OF PAPER

Parents who are concerned that lack of an official diploma will hurt their child needn't worry:

> *For the parents of high school age homeschoolers, getting a diploma is a big concern. Many new to homeschooling believe that the state will issue a diploma to them. As a private school, each homeschool issues their own diploma; however, I usually ask people what they want to use the diploma for and work back from there. For instance, if they have college in mind, I suggest that they contact the college and find out what is required for admittance. The same would apply for a job. I have not heard of any problems with this.*
>
> *—Susan, regional homeschool coordinator*

Contrary to popular thought, homeschoolers often don't need a high school diploma to go to college or enter the workforce, and they almost never need to take the GED (General Educational Development) test. Because the GED is often perceived to be primarily for high school dropouts, some homeschooling advocates advise against homeschoolers taking the GED test unless absolutely necessary. When seeking employment, homeschoolers can usually use their résumés to show their education and experience.

Homeschool families can grant their own diplomas. After all, a diploma is nothing more than a certificate recognizing completion of high school, and homeschooling parents, as the persons responsible for their children's education, can judge that their child is a homeschooled high school graduate.

Homeschoolers who want to have a high school diploma from an accredited institution can obtain one through an accredited umbrella school or a correspondence program. Although the terms umbrella school, correspondence school, and distance-learning program are sometimes used interchangeably, umbrella schools—also called cover schools—usually refer to businesses created to help homeschoolers with legal legitimacy. In many states, umbrella schools are unnecessary. In other states, such as Alabama and Florida, using an umbrella school is either one of the few options for homeschooling within the law or is by far the easiest option. These schools help parents to maintain attendance records, issue report cards and transcripts, and grant diplomas.

For college admission, more important than a diploma is a detailed transcript or description of a child's high school education. Remember that students from schools also don't have their diplomas when they apply to colleges; they are applying based on an education in progress. Homeschooled teens are wise to start keeping good records for themselves early in their high school years. These records can include descriptions of subjects they've studied, book lists, awards and accomplishments, sports, hobbies, projects, volunteer activities, jobs, and anything else that would be relevant on a college application. Some colleges ask for portfolios from homeschoolers, others for lists of subjects studied or specific college entrance subject exams. As Susan suggested above, students can contact potential colleges for specific information about requirements from homeschooled applicants. Several colleges' Web sites provide application information for prospective homeschooled students.

TWO HOMESCHOOL-TO-COLLEGE MYTHS

Two widespread myths have risen regarding college-bound homeschoolers. The first, left over from the days when homeschooling was largely unknown, is the myth that homeschoolers have a hard time getting into college. The second is more recent and stems from media stories about homeschoolers' admittance to and success in Ivy League and other colleges: the myth that homeschoolers can have their pick of universities.

The first myth is easier to correct. At least 95 percent of American colleges are open to homeschooled applicants.[1] While they may need to put some thought and work into translating a home education into the language of a high school transcript, homeschooled teens will face an even playing field when they submit their applications. If they have taken the task of being an educated person seriously, they will be as ready as any other high school student for college life. My experience as a college instructor is that students who have been homeschooled come to higher education with better than average self-motivation and purpose, work ethic, and love of learning. They are a joy to teach.

Regarding the second myth, homeschoolers should not think they will be a shoe-in for the college of their choice just because homeschooling sets them apart. Yes, homeschoolers can certainly go to college, but being able to go to college—even a very good college—is not the same as being able to go anywhere one wishes. This has nothing to do homeschooling and everything to do with the current climate of college competition. Michael Winerip, who interviews potential applicants for Harvard, wrote in the *New York Times* about the fierce competition for Ivy League admissions and the change he's seen since he went to college:

> What kind of kid doesn't get into Harvard? Well, there was the charming boy I interviewed with 1560 SATs. He did cancer research in the summer; played two instruments in three orchestras; and composed his own music. He redid the computer system for his student paper, loved to cook and was writing his own cook-

book. One of his specialties was snapper poached in tea and served with noodle cake.

At his age, when I got hungry, I made myself peanut butter and jam on white bread and got into Harvard.[2]

There are many good reasons to homeschool, but to do so primarily in hopes of getting into a top-tier school is not one of them. The good news is that well-prepared homeschoolers probably are just as competitive as anyone else, and they are in a good position to find schools that are right for them, even the best schools. The bad news—which, as we will see, might not be so bad after all—is that as long as the population of high school graduates continues to grow and each student on average continues to apply to a greater number of colleges (the Common Application has contributed to this trend), admission rates will continue to go down and rejection letters will fill high school seniors' mailboxes.

There is an upside to this fierce competition for college admission. As a society, we are finally realizing that getting a quality education is less a matter of the name on the school than the student's effort, something that homeschoolers have known for a long time. The best school in terms of rankings might be the worst fit for an individual student, who could get just as good an education elsewhere. Homeschoolers who have learned how to have self-direction and purpose in their learning can make sure they get a good college education almost anywhere they go.

TEACHING THEMSELVES

So, we know that homeschoolers can go to college, but the question remains: How are homeschooling parents able to teach their children algebra, trigonometry, chemistry, or physics? Shakespeare and essay writing? The answer is simple: They often don't. The teenagers teach themselves.

If this sounds strange, consider for a moment what really happens in many high school classes. Students are assigned reading from a textbook. The teacher goes over the reading, answers questions, initiates discussion, and assigns homework. Students do the homework and hand it in. At the end of a unit or term, they take a test. Class time is filled with some combination of lecture and discussion, group activities, taking attendance and other routine matters, and addressing students' questions. Most of the actual learning consists of student work: reading, understanding, doing homework, taking tests. Much of what the teacher does—assigning and grading, in particular—could be handled by someone else, even the students themselves.

That's not to say that teachers do nothing of importance, just that homeschoolers can find other ways to learn what is typically provided by a classroom teacher. The methods and resources that work best depend on the student's learning preferences, abilities, and other needs. Homeschooled teens often manage their own course of study in one or all subjects, using guides and books to assign themselves reading and problems, checking their own work or asking someone else to check it for them, and moving on when they have a firm understanding of the material.

The following are just a few examples of how homeschooled teens learn what they need to know, either on their own or with an adult teacher or facilitator.

Self-Teaching Guides and Materials Designed for Homeschoolers

Browse any large bookstore or Amazon.com and you will find self-teaching guides for almost any topic you can imagine: math, from pre-algebra through calculus; all science subjects; computer programming languages and Web design; grammar and writing skills; even philosophy and literature. Some of these books are meant for adult lifelong learners. Other study guides offer preparation for specific exams, such as the SAT or Advanced Placement (AP) tests. And some materials are written to help students to learn prerequisite material for harder courses.

The advantage for homeschoolers is that all of these resources are created for learners to use on their own, not for teachers to use in a classroom. They are written with self-direction and lifelong learning in mind, which affects the tone and presentation of the material. A good example of the best of what self-teaching guides can offer is Barron's *Easy Way* series, which often explains difficult or dry subject matter through stories and metaphor, and uses a friendly, conversational tone telling the reader what he or she can expect to learn and how to learn it. Other popular series include the McGraw-Hill *Teach Yourself* series and the Wiley *Self-Teaching Guide* series. These books are available on a wide range of topics and are available in many libraries and most large bookstores.

In addition to books, homeschooled teens use software programs designed to teach themselves about specific topics, especially foreign languages. Rosetta Stone and Power-Glide are two examples of popular foreign language programs that allow students to learn on their own. My experience is that software designed for adults who are returning to education is often much better—more interesting and engaging—than software meant to cover high school subjects or to help teens do well on standardized tests.

Teenagers who don't like this much self-direction may find teach-yourself guides frustrating rather than liberating. These materials can be of extremely high quality; some, however, skip over important information or present material in a haphazard and unconnected fashion. For students who want more outside direction or a more detailed structure, community classes, distance-learning courses, or individual study with adult direction may be better options.

Distance-Learning Courses and Electronic Learning (E-learning)

The growth of distance learning and e-learning has opened numerous possibilities for homeschoolers—especially teenagers—to embark on courses of study otherwise difficult to find. Our son has taken several high school distance-learning courses in areas of his interest with which his dad and I are unfamiliar, such as classic science fiction, medieval history, and Japanese. For each course, he used a print study guide and textbooks, submitted progress evaluations online, and took midterm and final exams under the proctorship of our local children's librarian.

Some distance-learning programs are designed for homeschoolers, while others also serve children in school who want to supplement their classroom education. Some offer one-on-one contact with teachers, either through email or chat rooms, and the teachers offer written feedback on writing and other work. Other courses are computer graded and assume the students will go through the material on their own.

Gene Maeroff writes in *A Classroom of One: How Online Learning Is Changing Our Schools and Colleges* that it's in precollegiate studies that homeschoolers benefit most from the growth of e-learning options.[3] High schoolers at home can use these courses to start high school studies early, to learn higher-level subjects that their parents can't teach or that are unusual, to do concurrent high school–and college-level work, and to earn high school or college credit.

Parents I've talked to report mixed success with e-learning. When it works, it works very well. It can encourage independent study and help teens to learn study skills and time management. But it's not a good fit for every child, and the quality of distance learning varies widely. E-learning can also be expensive, so it pays to ask others about which programs they have used and to do some research before paying for courses that may not be right for your child. Parents can research the Internet for specific distance-learning programs and electronic-learning Web sites. Note that using this option as a homeschooler to learn some subjects is different from enrolling in a virtual school as a full-time student. Virtual schools are often part of state public school systems, and the students enrolled in them are public school students. Other virtual learning programs are privately run. Parents need to decide for themselves if using one of these options for all of homeschooling is a good idea. In some states, such as Florida, using a private umbrella school—a distance learning or other institution that allows students to enroll and receive educational credit—is the only way to homeschool legally without reporting student progress to the state.

Mentors, Tutors, and Homeschool Co-ops

One of the most exciting ways that homeschooled children learn is by working with mentors and tutors, both formally and informally. Mentors or tutors can be older children, neighbors, relatives, co-workers—anyone who has knowledge in and passion for your child's interests. Some of the most important relationships our teenage son has had are with adults who aren't afraid to show that they still love to learn and are willing to share that love with young people: skating coaches and choreographers, children's theater instructors and directors, parents in our homeschool support group, music teachers.

Tutors usually focus on imparting specific knowledge and skills—how to write an essay or understand geometry, for example—while mentors serve as models of what it is like to have a certain job or hobby, to live according to certain values, or just to be

lifelong learners. Tutors can also be mentors, and mentors are sometimes tutors. Some tutors gradually become mentors as they form a
deeper relationship with the child and take on a larger role in the
child's overall education. David Albert calls this approach to homeschooling "community-based education," and in *And the Skylark
Sings with Me* he shares how community-based learning allowed his
daughters to learn what he and his partner could not teach them.[4]

Some homeschool families form homeschool co-ops (or less
formal support groups) to pool tutoring and mentoring services
for their children. These co-ops can be as small as just a handful of
families to large enough to require extensive planning and sched-
uling. For several years our family has been involved with a literature and writing group that meets weekly in families' homes.
Parents in our group volunteer to lead discussions and suggest assignments based on their interests and expertise. One parent who
is an acting professor leads the children in a playwriting workshop
and directs an annual Shakespeare production. Last year, a parent
with a law degree offered a unit on writing Supreme Court opinions, and two other parents teamed up to lead the children
through the reading and writing of poetry. In our group, parents
donate their time, but in larger and more formal co-ops, families
may pay fees for renting space to meet or to bring in outside teachers for certain subjects.

In Hadley, Massachusetts, North Star: Self-Directed Learning
for Teens takes the idea of homeschool co-ops even further. North
Star is a community center that offers teens homeschooling support,
guidance, space to socialize, and classes and other activities. Classes,
which are all voluntary and designed around the interests and needs
of teens and instructors, are taught by local teachers, parents, and
North Star staff. The adults at North Star serve as guides and facili-
tators for the teens, helping them to tap into their interests and improve their self-direction.

Adult mentors, whether one-on-one or in small groups, serve an important role for teenagers. Such relationships are part of healthy socialization, of learning to find one's place in the adult world. For homeschoolers, finding and using these mentors is much easier than for children in school, in part because their schedules are more flexible, and also because they have the time to devote to individual interests and abilities.

Conventional Textbooks

Homeschoolers tend to be suspicious of school textbooks, and with good reason. Textbooks are meant to be used by teachers in full classrooms, not by students on their own. They are designed for groups of children with diverse reading and ability levels, and, especially in the past several years, they are written to conform to state curriculum standards. These new standards-based textbooks are filled to the brim with factoids that ensure the books cover everything they are supposed to.

That said, while school textbooks have their drawbacks, they also can be extremely useful in certain situations and for certain homeschooled teens. The best textbooks are time proven, efficient, well written, and clear enough for a student to follow on his or her own. Those published for homeschoolers may have less of a classroom feel to them, and some textbooks meant for schools are also useful for individual learners. Publishers such as Saxon Publishers and Critical Thinking Company, which have proven track records with homeschoolers, offer catalogs and information tailored for homeschooling families.

Examples of excellent science and math texts used by homeschooling teens as well as classroom teachers are Paul Hewitt's *Conceptual Physics* (HarperCollins) and Harold Jacobs's *Mathematics: A Human Endeavor* (W. H. Freeman). I know homeschooling families who bought new editions of these books and others who used older editions from libraries or used bookstores. The first editions of these books were written before the standards movement, and later editions preserve the conversational tone and engaging explanations.

Textbooks can be as stifling or as useful as we allow them to be. The worst textbook can allow a highly motivated student to learn at least something, while the best textbook is useless in the hands of students who don't want to learn or believe they can't learn. Textbooks, like all other learning resources, are simply tools to be used or put aside.

Life, the Internet, and Everything: Self-Directed Learning

Cafi Cohen, author of *Homeschooling: The Teen Years,* found that about one-third of the families she surveyed moved toward less-structured learning for their high school years at home, even though those are the years when teens are planning for college and work.[5] Not only can self-directed learning work for high school, but when it works well, it is a joy to see.

Even teens whose education has been entirely self-directed have been able to document their learning on transcripts and get into college. In a *Home Education Magazine* article, "Fly-Fishing to College," Alison McKee describes her son's journey to college after he had spent much of his high school years obsessed with fly-fishing and fly-tying:

> [Our family] agreed that Christian's best chance at cracking the admissions process was to emphasize his unusual educational history. His competitive edge would be that he was unique, having no grades or traditional course work and vast experience living and learning in our community, in other states, and even in Germany. And, as it turned out, this was the case. The feedback from

college officials was that his uniqueness is what made him such a strong candidate.[6]

To teach themselves, these teenagers simply follow their curiosity wherever they can find answers. Just as when they were young children, they learn from reading, observing, thinking, asking questions, and engaging in conversations. In *The Teenage Liberation Handbook: How To Quit School and Get a Real Life and Education,* Grace Llewellyn shows teenagers the power they have to learn whatever they need and want to learn. High schoolers use self-directed learning in much the same ways that younger children do, but teens have more skills, resources, and freedom of movement. Their learning can look very structured from the outside, but because they are doing what they want to do for their own purposes, the structure comes from inside.

Karin is a fourteen-year-old who has directed her own learning all her life. Why does she homeschool?

> One, it's what I've always done. Two, I don't fancy the alternative. Three, I find I learn things faster than I would in school because I can go at my own pace. Four, when I get interested in something, then I learn about it, and I can skip over the parts that are easy for me and spend more time on the hard things. . . . I like that there is no set of things I have to do each day. Say, if I didn't want to do math today, I wouldn't have to, or if I wanted to spend the whole day out in the garden, I could.

Teens like Karin and Christian often follow educational and career paths that are different from the norm, such as starting an informal "major" by focusing on strong interests in high school, designing a high school education that will take them directly into the workforce, extending or shortening high school study, taking a year off before college, earning college credits through CLEP tests, or taking college classes while in high school.

SPORTS, PROM, AND ALL THAT JAZZ

I have talked to many homeschooled teens, and never have I heard anyone say they wished they could go to prom. Sports, however, are a different matter:

> There was really only one time that I seriously considered going to a school other than homeschool—in eighth grade, when I learned that it was not possible for me to play football if I did not attend high school full time. I made a serious choice at that point in my life, choosing what I perceived to be the greater benefits of staying home-schooled to the possible life path of football. I will never regret this choice. (Even if it means I will never be the next Brian Urlacher or Jerry Rice).   —Orlando

> There are some things that are a little more difficult to participate in because I'm a homeschooler. An example is playing a sport. It is nearly impossible to play on a school's team, and homeschooled teams are very often strongly religious based, which can be exclusive or contro-versial.   —Yasmin

For young athletes, homeschooling can be a help or a hin-drance. For some individual sports that are not connected to schools, such as figure skating, the flexibility of homeschooling makes it easier to train and maybe even cheaper, if they can use facil-ities and coaches at off-peak times. For other teens, continuing their involvement in team sports while homeschooling might take some compromise and ingenuity.

Many people think the first option is for homeschoolers to try to participate in public school sports. This doesn't often work, for a cou-ple of reasons. First, many if not most states and school districts do not allow homeschoolers to use school programs. Second, several homeschool advocates caution against involvement in public schools, since that could make homeschoolers accountable to the state and local government and could compromise many of the freedoms that homeschoolers worked hard to get.

But what are the alternatives? Some children do decide to go to high school mainly to stay competitive as student athletes. While they are giving up the flexibility of homeschooling, they are making a choice based on their own priorities. The experience of home-schooling is never lost or wasted, and these children can return to or start school knowing what is important and what is not important in their education.

Some children want to be involved in sports, but homeschool-ing is a higher priority for them. These homeschoolers can look for community leagues, teams sponsored by gyms and youth clubs, and church or civic sports program. Recent years have even seen the growth of homeschool sports teams and leagues, some of which have interscholastic competitions. If there is no homeschool team or league nearby, families can get together and start their own.

Homeschoolers have similar experiences with other activities that are provided by schools. Of the homeschooled teens I know, several are involved in music at many different levels, from jazz band to a city youth orchestra. Others are in church choirs, children's the-ater, community gymnastics organizations, and figure skating clubs. In some ways, finding social outlets for teens is easier than for younger children, because they can often form their own carpools and can take more responsibility for their own social calendar.

HOMESCHOOLERS IN THE CLASSROOM

During the high school years, some homeschoolers venture into the classroom—as homeschoolers. They do this by taking one or two high school classes or extracurricular activities (in states and school districts that permit it), as Tanya describes:

When I was fourteen, I considered attending high school for my senior year. I took two classes [as a freshman that year]—French II and choir—and then, in the next few years, art and French III. I have since decided not to go to high school full time ever. The high school environment is full of peer pressure, too many kids who just don't care and even a few teachers that seem to want the students to fail. I choose to homeschool among people who want me to succeed, who are so interested in learning that we cannot decide what to do first, and who will never pressure me into doing something stupid.

Homeschooled teens also sometimes take community college classes or attend university part time. Homeschooling is a very efficient way to learn, so it is not uncommon for teens to be ready for post-high school education before they have officially graduated from high school. Home education gives them the flexibility to make this transition from the safe base of home, where they can continue to become adults at their own pace.

HOMESCHOOLED TEENS: SOME FINAL THOUGHTS

I've been fortunate to know and work with several homeschooled teens, and I have a profound appreciation for their friendship and all they have taught me. They helped to inspire this book, and their example was one reason we continued to homeschool. What are homeschooled teens really like? They are as normal as any other teenagers and at the same time different in some important ways. Like all adolescents, they change and grow before our eyes. They can have mood shifts and emotional peaks and valleys. They have periods of extreme energy and stubborn lethargy. They think a lot about their future. They enjoy their friends. They begin to see with alarming clarity that their parents are far from perfect. They are concerned with self-identity. They sometimes think they understand more than they do, and they often understand far more than we think they do.

On the other hand, the homeschooled teens I know also resist many adolescent stereotypes. They are comfortable and candid in their conversations with adults. They have true friendships with the adults in their lives while simultaneously choosing to cooperate with adult authority. They belie the idea that parents and children can't be friends. They are accepting of the differences of other children, perhaps because homeschooling has given them a sense of what it is like to feel out of the mainstream. They are comfortable with self-directed study—some more than others—and they know how to ask for help when they need it. They don't focus much on grades, and they can be jealous of the free, unstructured time they have to follow individual interests and pursuits.

But enough of what I think. Presented below are what some of them have to say about their homeschooling, their future, and their families and friends.

TEENS TALK

What I Like about Homeschooling

Everything! I enjoy being able to work at my own pace for example. If I want to spend an entire day writing an essay for lit class and working on a paper for world history, I can! I also enjoy the independence. If there is a really interesting subject that I want to study—paranormal psychology, for instance—I can check books out of the library and spend the time I need to gather the information I want. Oh! And don't let me forget my friends! We are very close and there is NO peer pressure whatsoever, so our learning environment feels wonderful. Our literature class is only eighteen people and the parents who run it are all outstandingly talented people. We have a college professor, a former lawyer, an engineer, an acting professor and director, and an English major who are all dedicated to us. They teach us but also hang back and let us teach ourselves; instead of just telling us, "This is the way it is!" We are able to have long and intricate discussions about pretty much anything.   *—Tanya*

I have received the best education possible, an education that's gotten me ready for the next phase in my life in absolutely incredible ways. I've experienced things that no public school kid will ever be able to experience, been taught by incredibly talented people, and have been a part of the growth of a group of people that loves each person for who they are.   *—Orlando*

Being a homeschooler, I don't get as much academic stress and peer pressure.   *—Alyssa*

I liked the freedom to study what I found interesting. I missed that once I went to college, I felt part of me slowly dwindle away, since I lacked the time I used to be able to follow my curiosity with. However, I was better prepared than many of my peers in college for the different levels of work and the different kinds of people we'd be meeting there. I like the fact that homeschooling taught me to love learning.   *—Tim*

Goals for the Future

My personal goal as of this moment is to get to college and strive there to find my calling. I am accepted into Lawrence University for this coming spring, and am preparing for that experience.... Right now, I would dearly love to be a writer of fiction. I would even more dearly love to be considered a good writer of fiction. (It would not even be so bad to be a successful writer). But I would also love to be an actor, and a game designer, and maybe even a concert pianist. Right now, my deepest personal goal is to love the people I care about, and to continue to strive to find the path that's right for me.   *—Orlando*

At the moment my highest goal is to score well on the SAT and to get into one of the colleges I am interested in. Next year I would like to get a job and possibly study abroad in France.   —*Tanya*

Socialization

I take band from the public school. There is a community choir that my mom, my sister, and I are involved with. We also are in the church choir. I go to our homeschooling group and a writing class run by the parents of the kids in the homeschooling group. I take trumpet lessons, and I am on a homeschooling soccer team.   —*Karin*

I have participated in the art of Tae Kwon Do for the past six years. I have been involved in Girl Scouts for nine years, and through Girl Scouts I have traveled to many places, studied some very interesting things, and consider much of what I have learned through Girl Scouting an important part of my education. I'm also involved in the Youth Leadership Council through Girl Scouts, which involves creating and leading programs for younger girls as well as our peers, helping adults learn how to effectively communicate with teenagers, and learning how to make the world a better place, starting with ourselves. I have found these two outside-the-home activities (Tae Kwon Do and Girl Scouts) especially beneficial to my current and long term education. I am also involved in Shakespeare Home Players, a writing/literature class, I have a job working as a shelver at the library. I play on a homeschool girls' soccer team, and I also participate in many community service projects.   —*Yasmin*

THE HOMESCHOOLING DECISION:
HOMESCHOOLING FOR HIGH SCHOOL

Creating an individualized home education for teens can be exciting, fun, and a little scary. To get an idea of what your family's high school education at home might look like, you can borrow these three core considerations from North Star: Self-Directed Learning for Teens:[7]

- For Teenagers: What are your aspirations? What do you want to do? What are your short- and long-term goals? What are your areas of brilliance and struggle?
- For Parents: What do you think is important for your child to learn? What are your family's guidelines and values?
- For Teens and Parents: What is your state homeschool law? What plan do you have to homeschool within that law?

By talking about these three factors, families can work together to see if homeschooling for high school can meet everyone's needs.

DOING YOUR HOMEWORK:
SELECTED READING

From Homeschool to College and Work, by Allison McKee (Bittersweet House, 1998)

Homeschoolers' College Admissions Handbook, by Cafi Cohen (Prima Publishing, 2002)

Homeschooling: The Teen Years, by Cafi Cohen (Prima Publishing, 2000)

Homeschooling for Excellence, by David and Micki Colfax (Warner Books, 1998)

Real Lives: Eleven Teenagers Who Don't Go to School, by Llewellyn, Grace (Lowry House, 1993)

A Sense of Self: Listening to Homeschooled Adolescent Girls, by Susannah Sheffer (Boyton, 1997)

The Teenage Liberation Handbook: How to Quit School and Get a Real Life and Education, by Grace Llewellyn (Lowry House, 1998)

7

PARENT COFFEE BREAK

Frequently Asked Questions

Homeschoolers know that learning has its own rhythm, and sometimes the best thing to do when we feel overwhelmed or unsure of what to do next is to take a break. Let's take a break, one parent with another. Like most homeschooling parents, I am used to hearing a lot of questions. Sometimes we don't have the time to answer or we don't know how to answer thoroughly. Other times we're unsure of how our answers will be received. Most of the time, we are left feeling we didn't have a chance to give a good description of what homeschooling is really like.

This chapter is a chance to take all the time we need to answer the questions you might still have. I will not pretend to be unbiased about homeschooling. Home education is one of the best things that has happened to our family, and it's hard not to want others to know how fulfilling being a homeschool parent can be. At the same time, homeschooling is a personal choice, and it is not the best option for everyone. So, while my enthusiasm may occasionally show, I will also answer honestly, based on my homeschooling experience and at times sharing the experience of other parents. Then you can turn to chapter eight and decide if homeschooling might be in your family's future.

Pull up a chair, grab your favorite cuppa, and let's chat.

How do you know that you are qualified to teach your child?

Parents with no teaching background often worry that they are not qualified to be homeschooling parents. It doesn't help when family members or friends question our ability to teach our own children. Debbie described her early fears, and how they were resolved:

> *I didn't think I had the patience and I didn't think I had the smarts. I have learned patience (and am so grateful because it reduces stress in a big way!). I have also had the enormous and incredible support (day in and day out, mind you) of a very dear friend who at every turn was my cheerleader, reassuring me that my way is a fine way. She helped me to have the confidence to know that how I waded through this with my children was "all good." She also wanted our family to have success.*

Whatever patience and "smarts" Debbie felt she didn't have at the beginning, she found within herself or learned along the way. All parents—homeschooling or not—should have such a cheerleader in their corner, someone who supports their decisions and wants them to succeed. Those who don't, however, can think through the idea of qualifications, teacher certifications, and parenting before they convince themselves they don't have what it takes to homeschool.

First, keep in mind the distinction between having certification on the one hand and being qualified in the sense of being capable and competent on the other. Teaching certificates are issued by individual states or the National Board for Professional Teaching Standards. To get a teaching certificate, one usually needs a bachelor's degree as well as additional course work and testing. While public schools require teachers to be certified, many private schools do not, recognizing that certification does not guarantee high teaching quality. Some certified teachers are excellent in the classroom. Others are not. And some excellent teachers are not certified.

In recent decades, most states have adopted laws that do not require teaching certificates or any other formal qualifications for homeschooling parents. A few give the option of having a teaching certificate or passing a teaching exam as one way to homeschool legally, but they also give other ways to qualify according to the law. Legally, a teaching certificate is not necessary.

To decide if they are qualified to homeschool, parents can ask themselves if they are capable and competent parents, and what it means to be a good parent. When we have children, we are given the responsibility of caring for them and helping them to grow into healthy and responsible adults. While we may not think about it much, that's a big responsibility, one that we begin with no certification and little or no training. Our qualification in essence comes from the responsibility of being a parent, and we must choose for ourselves how seriously we take that responsibility and what it means to us.

Consider everything we do for our children before they reach school age: When they are infants, we tend to all their needs for the many months when they can do nothing for themselves. We are responsible for everything they need to keep them alive and well. How many of us initially feel qualified for this task? Yet we can do it. We learn as we go by paying attention to our children and learning from others. When they are babies, we can't imagine how we'll ever parent a toddler, but once they can move around, we help them to walk, to talk, to feed themselves and care for their needs. We teach them how to tie their shoes. We help them learn how to get along with others, how to share, how to stand up for themselves. Sometimes we teach them how to count or read, or we share our interests and hobbies. We show them how to follow their own interests as well, how to find information in libraries and take nature walks and visit museums. When they tell us everything they know about dinosaurs or bugs, we find ourselves learning from them and trying to keep up. We do all this without any certification, because it is our responsibility to our children.

Homeschooling is no different. We already have several years of experience teaching our children, and we know them better than anyone else. If they have been in school, we've helped them with homework. We've read to them and answered questions. In home-schooling, we continue the same process. We read books together and discuss them. We find interesting ways to learn about geometry. We learn about the countries of the world. We find classes and groups to be involved in. We love them. We are, in short, good parents. Not perfect parents or perfect families. Just good parents whose children are educated at home.

If you are a capable and competent parent, you are qualified to homeschool your children.

Do homeschoolers follow a national or state curriculum?

The United States, unlike some other countries (the United Kingdom, for example), does not have a national curriculum. The tradition in the United States has been for education to be controlled at the state level and by individual school districts, to better serve local needs. While some organizations such as the National Council of Teachers of Mathematics and the National Science Teachers Association publish and promote their own suggested national standards, such guidelines are not agreed upon by everyone, and each state ultimately chooses what its students will be required to study. So home-schoolers in the United States do not follow a national curriculum.

Each state adopts its own public school curriculum, and many states require homeschoolers to study the same subjects as public school students, although other states require no specific subject areas. State curriculum specifications for homeschoolers vary from a simple listing of subjects (English, math, science, and so on), to more extensive curricula for each grade, to requiring the submission of a detailed curriculum outline to local school officials. If by state curriculum you mean the state prescribing a year's worth of books and lessons for homeschool families to follow on a week-by-week basis, then the answer is also no.

A good point of comparison is the state public virtual school. These virtual schools are part of the public school system, and in most cases, the district that hosts the school receives funds from the state for each student enrolled. Virtual school students receive a cur-riculum package, which might include software, books, access to online materials, a computer, and other educational support. While the parents oversee the children's study at home, the state is ulti-mately in charge of the child's education. This relatively new form of public schooling may benefit certain children and families, but it is not the same as homeschooling, which allows for a more individual-ized education and places control and accountability in the hands of parents.

Parents new to homeschooling might think that it's a good idea to follow their state's public school curriculum standards and course of study, even if they are not legally required to do so. However, while state education requirements are interesting and offer insight into the educational process, they are not necessarily the best model for homeschooling. Standardized curricula are divorced from indi-vidual children's abilities, interests, needs, and circumstances, and they often reflect the political and economic concerns of school rankings and textbook publishing. Susan offers this explanation to new homeschoolers:

> *One of the most frequent concerns is curriculum: What should I use? Where do I get it? I let them know that there is no set curriculum that is required or provided. In the state of Wisconsin, homeschools are considered private schools and each chooses their own curriculum in keeping with the law which requires 875 hours per year of a "sequentially progressive curriculum of fundamental instruction in reading, language arts, mathematics, social studies, science and health." Some people have a hard time understanding this because we are so used to the school model where we are told what we need to cover in every grade. In reality, every individual needs different knowledge and skills to competently handle their life, and it would be impossible to provide all of this knowledge, whether in or out of a school. Homeschoolers often put a high priority on lifelong learning and come to know how they learn best. This enables them to gain the understandings that they need as life moves forward.*

Homeschooling teens who do not want to go to college can design their own high school curriculum individualized to their interests, talents, and goals. They can create their own unique course of study, without the restrictions of academic tracks or unnecessary curricular requirements. They can focus on areas of strength without having to fulfill someone's idea of being well rounded. They can choose to forgo any standardized tests not required by their state laws. They know that testing is not as important an indicator of their progress as their own sense of what they learn and understand.

But without standards, how do parents know what to teach?

Here's where the fun starts for parents! Getting involved in your child's education and deciding together what to learn is an extremely rewarding experience. If you don't believe me, sit down with your children and tell them that this year they will have a say in what they study. Ask them to put together a list of subjects, topics, skills, and questions that they want to pursue as part of their homeschooling. Then watch their eyes light up.

Because state laws and homeschooling approaches vary so widely, each child's homeschool studies will be different. Here are just a few ways that families decide what to study or learn:

- Parents who want to include the conventional subjects can ask their children what aspect of each subject they'd like to learn about. Almost any topic can be adjusted to the child's grade level. An eight-year-old who wants to learn about the stars will approach science differently from a fifteen-year-old who is ready to learn astronomy and physics. Giving the children a choice of how to approach each subject makes it more likely that they will be motivated to learn. They can also help with finding resources.

- Children and parents together can look at the World Book Typical Course of Study Web site to see what is typically taught at each grade level in school classrooms.[1] These lists can be used for ideas or as a general curriculum outline. Homeschooling allows families to think "outside the grade," so feel free to look at lists from several grade levels to put together an individualized course of study.

- When our son wanted to learn about a specific topic and I wasn't sure where to start, I sometimes found course descriptions from distance-learning programs and even school Web sites to get ideas for books to read, activities to try, and questions to ask. These course listings can also give you ideas about what subjects you might want to "offer" your homeschooled child. Taking the idea a step further, your child might enjoy being given a list of "available courses" that he or she can "sign up for."

- Families can let their children's interests and goals determine what to study. A child who wants to plant a garden can do so, and will learn how to prepare the soil, choose the seeds, and pull the weeds. A child whose goal is to read a biography of every United States president might spend months doing just that. Not all homeschoolers have intensive goals. Sometimes they follow a train of questions and interests that, at the end of the day, adds up to more education than they would have gotten with a more formal curriculum.

- Families who want to provide a more conventional education can choose to buy curriculum packages for specific grades from either traditional educational publishers or homeschool companies. This option eliminates the need to decide what to study, because the subjects are chosen by the publisher; at the same time, this option leaves little room for meeting individual interests. Parents can involve their children in this decision by researching different options together and, if possible, borrowing samples from friends to try out. When the box arrives, the children can unpack their new books and study guides, excited to begin a new year.

- Rather than follow a grade-by-grade course of study, families can discuss what subjects are usually learned in elementary school, middle school, and high school, then use that framework to decide on more specific topics. This approach allows for the flexibility necessary to meet individual needs, while giving parents some reassurance that their homeschooling has direction, as Debbie explains: "I have no formal curriculum. We do parallel *very loosely* the subjects that kids follow in public or private schools. It is progressive. We do keep track of our child's grade level mostly so that when he is out in the world and people ask him what grade he's in, he has an easy answer."

- As children get older, they tend to choose what to study based on their long-term goals. A child who knows she wants to attend a specific college will follow a course of study that meets that college's application requirements. A different child who plans to get a computer job after high school will teach himself as many programming languages as he can and look for IT jobs or internships. Surprisingly, because they are working toward a goal, choosing what to study is sometimes easier for high school–age homeschoolers than for younger children.

One of the most surprising revelations for parents who homeschool is that there really is no one course of study. The universal curriculum is a myth, left over from a time when it might have been possible to teach all children in a small school a unified body of knowledge and set of skills that most adults agreed on. As our world grows more complicated and we have access to more information than we ever thought possible, the idea of ticking off the same list of learning outcomes for every student becomes not only less practical but more absurd. Every time we add a new required subject or topic to the curriculum—whether it's computer skills or interpersonal communications—something else has got to be taken off or everything needs to be watered down. Some children have little interest in or need for much of what they must learn to pass standardized tests, and others already know the material and could use their time learning something new. Children who are homeschooled can focus their studies on what they need and want to learn, not what a faraway committee has decided.

*Isn't it unhealthy for children to spend
so much time with their parents?*

This question always makes me a little sad.

One of the unfortunate consequences of spending most of their days apart from each other is that families come not to know how to spend time together, how to enjoy each other's company. Summer vacations are a good example. As the days get longer in the spring, so, too, do parents' worries and complaints about how to keep their kids busy in the summer without anyone going crazy. Summer parenting magazines are filled with advice about how to get children "off the couch" or away from video games, how to find summer camps to fill up some of the time, and how to make the countdown until fall a little easier. The rare parent who does admit to looking forward to three months of togetherness may be viewed as unnatural at best, clingy or controlling at worst.

While it may be unusual for parents and children in the twenty-first century to spend most of their time together, it is far from unhealthy. There is nothing wrong with a family whose members respect each other, can spend long stretches of time with each other comfortably, and genuinely enjoy that time together. How many parents *wouldn't* want that kind of relationship with their children? How many families have the time to practice such a relationship?

Families coming to homeschooling from many years in school sometimes are uncomfortable with being together so much, at least at the beginning. In time, they learn how to be a family again, not just in name, but in practice. Watch any veteran homeschooling family to see how this works. Parents learn that they can be with their children without needing to comment on every action or word. Children learn that their parents are capable of interesting conversations and even a little fun.

The value of having more time with their mothers and fathers is not lost on homeschooled children. When asked about his long-term goals, Tim, a former homeschooler who now is a project manager at a software company, says, "I hope to get married and have children. Oh, and I want to homeschool my kids. Be a stay at home dad and raise them."

Is homeschooling expensive?

No, yes, and maybe.

No, it doesn't have to be. Many parents homeschool for next to nothing. Any books and toys they buy and community classes they sign up for would have been purchased even if their children were in school. They borrow materials from the library or friends, and they barter with other families for music lessons, mentoring, or babysitting. There is no need for a school uniform or school activity fees, and certainly no school tuition. Many museums, bookstores, and teacher supply stores offer discounts to homeschoolers. And the Internet is a wealth of free resources, from reading guides and lesson plans to online textbooks and classes. Especially for children whose needs are not unusual, homeschooling is a very cost-effective education. Kathleen agrees: "Homeschooling is inexpensive because we've always made good use of several area libraries and inter-library loans. We purchase some new materials most years, but also re-use, borrow, and lend and share lots of games, texts, and other learning materials." Kathleen mentioned a math textbook that her youngest child (of five) has started using. The inside cover has a "This book is the property of" stamp, and she saw the names of not only two of her other children, but three other homeschoolers she knows, and now her young son. The homeschooling history penciled in the book made her smile.

For other families, yes, homeschooling is expensive. Children with unusual learning needs may require costly resources that aren't found in libraries. Homeschooling families pay for therapies and special classes that schools would otherwise provide. Families forgo a second income so they can homeschool. Families who use umbrella schools or private tutors to fulfill state requirements must pay for these services out of pocket, and for large families, the costs for each child add up fast.

Becca explains how homeschooling her oldest child costs more than homeschooling her middle son:

> *Homeschooling can be expensive but it doesn't have to be.... Homeschooling my middle son has been cheap! We use mostly library books and videos and an occasional curriculum. Homeschooling my oldest son has, at times, been very expensive because he's interested in so many things and I couldn't teach him most of these things myself. So, I've paid for tutors, group classes, private lessons, more expensive textbooks, equipment, and more. I've had the great privilege of being able to take advantage of several groups who help support highly gifted children in their endeavors so we've been able to manage the finances. It's still less expensive than private school, though, and well worth the expense.*

For most homeschoolers, homeschooling is as expensive as they want it to be. With few exceptions, parents who homeschool can decide how much they want to spend on home education, and stick to that figure. This doesn't mean that children can take advantage of every class or opportunity that is offered. However, homeschool parents are wise to remember that a good education never hinges on finding the perfect book or class. With planning and creativity, homeschooling can be made to fit almost any budget. Some distance-learning and other programs offer scholarships for homeschoolers, and grandparents or other family members may be willing to pay for homeschooling costs as an educational gift.

Can I continue working if we homeschool?

The answer to this question depends a lot on individual families and jobs. Some parents are willing and able to work different shifts so that they can homeschool. Others can bring their children with them to their jobs or can do work at home. In some cases, a parent puts a career on hold while homeschooled children are young. Partners can discuss if one of them is more interested in being a primary homeschool parent, or if they want to share homeschool duties.

For many modern families, juggling homeschooling and work—while sometimes difficult and stressful—offers more advantages and brings less stress than juggling school schedules and work, as Lisa Wood, who juggles freelance writing and homeschooling, explains: "I watch people whose kids go to school, and that's a lot too—hustling to get them out early to the bus, dealing with issues with the school. . . . Either way, educating a child is demanding."[2]

Parents can keep in mind that homeschooling is an efficient way to learn, and their children will not need to spend six or more hours per day in intensive study. Homeschooling also gives parents more control over their children's schedules, which can make it easier to plan for vacation time. If this is something you are considering, it's a good idea to have an honest conversation with your employer to know what options are available to you.

Can children who are not yet of school age homeschool?

Legally, children are not homeschoolers until they have reached the minimum age of compulsory attendance for their state. If, from a different perspective, homeschooling is thought of more as an approach to family life than an educational option, then many children are homeschooled from birth. In homeschool support groups, I have heard people question whether parents of two-, three-, or four-year-olds are really homeschooling. My opinion on this has changed through the years. At first, I thought, "No, they aren't really homeschooling until their children are school age." Now I see a difference between parents who assume their children will go to kindergarten and those who plan to homeschool.

The parents who plan to homeschool have a strong sense of their role in their children's education even when their children are very young. There is no thought of getting a preschooler ready for the kindergarten classroom or giving the child a head start of any kind. These parents are not preparing their children for the inevitable separation anxiety or the experience of getting on a bus alone. They display a self-confidence in their parenting and comfort with their children unusual for young parents. Many parents who do not plan to homeschool show these same traits, but I see them more often and more clearly in the parents who are planning to homeschool.

If you have a very young child and are thinking of homeschooling eventually, it's never too early to look for support groups or think about what your homeschooling will look like. The transition from preschool to homeschool will be that much easier.

Can I homeschool an only child?

This is a question I can answer from personal experience. Yes! Parents who homeschool an only child do have a few decisions to make, but first I want to say what the advantages are, because there are many.

The greatest advantage is that only children who homeschool have more opportunities to find surrogate siblings than if they were in a classroom. In school, an only child will learn and socialize with children his or her age. As homeschoolers, only children can socialize with children of all ages. Even the more structured homeschool classes and co-ops usually consist of age ranges rather than a specific age or grade. This allows only children to learn how to relate to older children and younger children. This year our son is good friends with children his own age as well as children a few years older, and he writes an ongoing Star Wars story via email with a homeschooling friend who is several years younger than he is. Once we started homeschooling, I saw his relationships with younger children improve and grow, as he had more of a chance to get to know them.

Because only children are used to being with adults, they often find the one-on-one nature of homeschooling a comfortable fit. Parents need to concern themselves with meeting the needs of only one child, making preparation relatively easy. Also, there is nearly limitless potential for engaging discussion and sharing of interests and knowledge between parent and child.

Parents who homeschool only children need to be willing to find or plan social activities and opportunities for friendships. If an only child is an extrovert, parents might spend quite a lot of time chauffeuring to friends' houses or hosting gatherings, while introverted only children may need some extra encouragement to seek companionship with other children. Parents can also take care to help only children learn that adults, too, need quiet and alone time occasionally.

*How do parents teach several children
at a time, and will siblings get along?*

Here are what some parents of more than one child have to say. Becca homeschools three boys, Debbie homeschooled two children until her daughter decided to go to school, and Kathleen explains how her son displayed homeschooling initiative to make sure he got his share of his mother's attention:

> *All my boys have periods where they get along well and periods where they squabble. I think homeschooling has benefited them a great deal.*

I believe they're closer to each other as friends than if they had gone to school all day. My older boys are very different in personality (an extreme introvert and a strong extrovert), but because we homeschool, I believe they've found more in common and they've learned to respect the other's interests because we've been able to dialogue about this at home. —Becca

As soon as we decided to homeschool our seven year old, our eleven year old said she wanted to try it. She did. She loved it. She decided to go back to public high school, though, after two years, and is now a freshman. She went back without a hitch. She is really well-adjusted and has an A and B average. Needless to say, my son and I miss her and her good energy around the house! —Debbie

Four school-age children plus an infant, one mom . . . you get the picture! It was brought to my attention (by my oldest child) that I was spending a lot of time with our daughters' Girl Scout troop. He suggested a solution—he started a Mom and Sons book discussion group that met monthly (like the Scouts). He had my full attention for an evening, surrounded by friends and their moms, talking about a book chosen by one of them —Kathleen

Parents who homeschool several children find ways to work with their children together, such as reading a book that everyone understands or learning about a topic that each child can approach at his or her own level. They also set aside times to work with each child individually while the other children play, do independent work, or are in activities outside the house. As children become used to homeschooling, their sibling problems often decrease, and they are able to work together more effectively. Some parents with several children are excellent family managers, and use well-thought-out schedules to ensure that everyone's day runs smoothly. Other parents juggle the needs of several children with intuition and creativity. Whatever parenting style works best is usually also the home-schooling approach that will work well.

What are some of the challenges of being a homeschooling parent?

Parenting comes with challenges, whether our children go to school or are homeschooled. The challenges of homeschooled parents are different from those of school parents, not necessarily easier or harder. Some parents, like Becca, face the challenge of managing the education and daily lives of several children:

> *One of the biggest challenges over the past two and one-half years has been homeschooling with three children. I have lots of friends who homeschool with more children than I have, and it amazes me that they do so well! My toddler is still so needy, and I don't spend much one on one time with my older sons unless my two year old is napping. So one of my goals in homeschooling is to have my children work more independently. . . . There are days when we simply can't do some activities until my husband comes home.*

Debbie described another common challenge: juggling home-schooling, a desire to work, and financial needs:

> *We are a two-income family. My husband works a nine-to-five job as a teacher. I am a massage therapist, so I have some flexibility with my schedule, but that means that I do a lot of juggling. My son gets the more formal schooling in the morning and the more experiential experiences in the afternoon. We struggle financially, so it is my inclination to work whenever I can. I do sometimes worry that it takes away from my son's schooling time with me.*

Homeschoolers must also deal with the challenge of explaining their choices to friends, family, and even strangers who question why children are not in school. Kathleen, who has homeschooled five children over a span of twenty-three years, shows how she used this challenge to her advantage:

> *The biggest challenges were answering all of the schooled world's questions about our decision to homeschool. I felt it was no one's business but ours, so I didn't enjoy the process of explaining to skeptics, although doing so often reaffirmed my goals, strengthened my resolve, and sometimes caused some healthy doubts that led to positive change.*

Another challenge that almost all homeschooling parents face is finding time for their own social needs and personal interests. Having a supportive partner helps a lot, as does helping children to understand that every family member has these same needs. However, the primary homeschooling parent often has to stand firm and insist on this time for himself or herself. Be sure not to lose contact with old friends. Keep up or find new hobbies. If life gets busy and something's got to give, don't let it be your own health or sanity. Instead, skip the cleaning once in a while or use paper plates.

What are the greatest joys of being a homeschooling parent?

Along with the challenges are the joys, and the joys of homeschooling are many. I can describe them no better than the parents I interviewed did:

> *My relationship with my children is deeper and we've had more time together to really get to know each other. We all really like each other. I have had an opportunity to learn the things I let go in one ear and out the other as a kid. I think my children are learning on a deeper level than I ever did in school. I believe this is knowledge that they want to learn and will stay with them. We are part of a homeschooling group that is unique: diverse, supportive, dynamic, open, the list goes on and*

on. It is really great to be a part of a group whose members all home-school for very different reasons and in many different ways. I also think it feels like the more "natural" way to learn. —Debbie

The joys of homeschooling are many! Watching my middle son become an avid and advanced reader has been wonderful. I've been able to guide him in his reading choices and watch his confidence bloom as he went from barely reading to reading anything in the span of six or so months. Another pleasure of homeschooling has been watching my children immerse themselves in their interests. My oldest son has always had so many passions and the freedom he's had to pursue these in depth is irreplaceable. My sons have gone down such interesting trails while learning. Feeding their hungry minds has been so much fun!

The wonderful conversations we've had have been a rich source of joy and learning. What a blessing it has been to have time to talk with my children in depth; I find we learn much from each other. One of our favorite activities is to recommend books to each other. I love sitting back and watching my kids learn! I love providing them with materials and opportunities and then seeing what transpires. I love watching them decide what to learn. My middle son decided he wanted to learn about ninjas and samurais this year and it was great tracking down loads of good books from the library for him to devour.

Homeschooling takes effort on the part of the parents and the child, but is a well worthwhile effort to make. Homeschooling has been an experience I would not trade for anything. —Becca

Greatest Joys . . . Hearing my kids thank me for homeschooling them (ten years before I wondered if they'd blame me for ruining their lives). Hearing of my children's plans to homeschool their future families. Attending the graduation of our eldest child who made his way to the school of his choice, earned the degree of his choice and had a lot of fun while doing it. Experiencing the wholeness of being so near and available while girls grow into women and boys become men. —Kathleen

I wish other people knew what a great adventure homeschooling is. It isn't easy, sometimes it's a little scary, but it's always an adventure. Through homeschooling, I have met some of the nicest people I know, I have learned invaluable life lessons, and I have become a better person myself.   *—Susan*

8

THE HOMESCHOOLING OPTION

Making the Decision and Getting Started

For many parents, homeschooling is a matter of heart and belief;
they feel deeply and strongly that they should be homeschooling.
They sometimes start homeschooling with only this conviction to
go on, unsure of exactly how they will proceed, and they work
out practical details and solve problems as they go along.

— *Larry and Susan Kaseman,*
"Taking Charge through Homeschooling"

MAKING THE DECISION

Is homeschooling a good option? Yes. We've seen that home education can provide families with everything they need to extend their
natural role of parents to their children's education.

The question of whether homeschooling is a good option for a
particular child or family is another issue, and one that only you can

answer. This chapter offers questions and ideas to guide you toward a decision. If you do decide to homeschool, you'll also get some suggestions for getting started.

And what if you decide not to homeschool after all? Give yourself a pat on the back for thinking the matter through carefully and making the best decision for you and your child! You may reconsider homeschooling at some point in the future. Even if you never homeschool, understanding what the homeschooling option is really like will enrich your understanding of education and family life.

There is no need to rush into a decision or to feel pressure to choose one way or another. Take your time. You will know what is right for you.

Know Your State's Homeschool Law

If you haven't done so already, make sure you understand what you are required to do to homeschool legally in your state (see the appendix for contact information). Some states have laws that are easier to grasp on a first reading than others. If you have questions about the law, get in touch with a state homeschool association or a local support group. For almost all homeschoolers, fulfilling the law is not a big deal if you are well informed and in touch with other homeschoolers who can answer your questions. You should know your state's homeschooling requirements for each of these areas:

- notification of intent to homeschool
- ages of compulsory attendance and cutoff birth dates
- qualifications of parents
- attendance requirements
- private school status or homeschool status
- subjects and curriculum to be covered

- contact with local school districts
- testing and assessment requirements
- record keeping
- special-needs provisions

Why Do You Want to Homeschool?

Make a list of what you hope to gain from home education. Think about these kinds of questions: What do you imagine your days will be like? What learning needs can homeschooling meet? If your children are in school, what do you hope will change? What specific problems do you expect homeschooling to solve?

To help you with these questions, here are just a few of the reasons that families choose homeschooling:

- Can give child better education at home
- Religious reasons
- Poor learning environment at school
- Family reasons
- To develop character/morality
- Object to what school teaches
- School does not challenge child
- Other problems with available schools
- Student behavior problems at school
- Child has special needs/disability[1]

Some of these reasons may be yours as well, or you may have very different reasons. It's no exaggeration to say that no two families homeschool for exactly the same reasons. If your reason is not on this list, that doesn't mean it's not a good reason to homeschool.

Becca, like many other homeschool parents, has found that her reasons for homeschooling are varied:

I began homeschooling because my oldest son was so active and so academically advanced that I knew he could never fit into a school system. My husband and I also realized how important our faith in Christ had become to us, and homeschooling allowed us to share this faith with our children in depth. Finally, my husband and I realized how fleeting life is; spending time with our children is such a privilege. Homeschooling allows us to share our lives with each other at a level we just couldn't have, had they gone to school. Lastly, we really don't want government schools, a recent phenomenon in the history of mankind, raising our children and dictating what they learn.

By delineating her reasons very clearly, Becca is able to structure her homeschooling to meet these needs. Parents who are already homeschooling can revisit their reasons every so often. Original goals or concerns can change as children grow older, and being aware of those changes allows families to adjust their homeschooling to fit their new purposes.

One more important thing to remember: You might have no other reason than "I want to" or "It feels right." That's good enough, too.

What Concerns or Fears Do You Have about Homeschooling?

Homeschooling is a scary idea for many people! Almost every homeschooler had to wrestle with these kinds of fears and concerns at the beginning (I know I did):

- Do I know enough to teach my child?
- What if my child won't respect me as a teacher?
- What will the neighbors say?
- What will my parents say?
- Will the child be able to go to college?

- Will we drive each other crazy cooped up here all day?
- What if homeschooling turns out to be a mistake?

Don't ignore these fears and hope they go away. Share them with members of your family or talk to other homeschoolers. Think about whether your fears are realistic. Often simply talking through our concerns puts them in a clearer light. All of the above thoughts, for example, stem from lack of self-confidence, concern over what others think of us, or unfounded worries about the future. Once families begin to homeschool, parents often gain self-confidence, learn to trust their own instincts and experience, and stop worrying about the imagined pitfalls of homeschooling.

If any fears continue to linger after you've talked them through, give some careful thought as to whether your concern is large enough, at least for the time being, that you do not want to homeschool. If you can articulate a clear and specific reason for not homeschooling, your decision will be much easier than if your decision is based only on vague doubts.

What Do Your Children Think about Homeschooling?

Ask them. Then listen. While the decision to homeschool and the responsibility for the child's education is ultimately the parent's, the child's opinion should also matter.

If your child is the one who wants to homeschool and you are unsure, spend some time talking about what he or she expects homeschooling to be like. Does your child know someone who homeschools and wants life to be like that? If so, would you homeschool in the same way, or would your homeschooling days look different? Often we assume we know what our children think, but when we sit down and really talk about issues that are important, we gain an entirely new understanding.

If your high school–age child wants to homeschool, talk about how much independent learning he or she wants to do, and how homeschooling fits with long-term goals. Together with your teen, put together a plan that clearly states what parts of his or her home education you are willing to oversee and what parts are the teen's responsibilities. See if you could make it work.

Maybe you are more amenable to homeschooling and your child is hesitant. Ask about his or her concerns or fears. Do your children know what homeschooling is really like? They might be basing their opinions on stereotypes or what they've heard others say or how one particular family homeschools. Or they may be fearful of what their friends or others might think. Give them books to read or take them to a homeschool support group or conference to give them more information. Or give homeschooling a trial period of a week or a month or the summer before making your decision. Keep in mind that many of the benefits of homeschooling take time to develop, but even this short period will give you a better sense of what it's like to learn at home.

Are you friends with a homeschooling family? Ask if you or your child can have a homeschool shadow day. High schools and colleges offer shadow days when prospective students can accompany a current student through the day. The "shadow" goes to classes, lunch, and other activities to get a better idea of what life in that school is like. There is no reason why prospective homeschoolers can't ask for a similar experience. This works best if you know a family whose approach to education and values are similar to your own; otherwise, keep in mind that your own homeschooling day might look very different. Even if the family's day is different from how you would homeschool, it will give you some ideas for your own home education.

If, after learning more about home education, your child is still firmly against homeschooling, give careful thought to how this will affect your day-to-day experience. Families do find ways to make homeschooling work when children don't want to learn at home, and often children themselves come to change their minds about homeschooling. However, it's always easier and more pleasant if children can feel a part of the decision-making process and feel they have at least had a say in reaching a compromise. If you decide you want to homeschool and your child is firmly against it, you might suggest a trial period, or you can work with your child to create a homeschooling experience that addresses some of his or her objections.

What Are Your Child's Social and Emotional Needs?

Use the following questions to think about your child's social and emotional needs, and whether homeschooling would help your child to grow toward self-knowledge, self-acceptance, tolerance and understanding of others, independence and interdependence, and fulfillment of personal potential.

- Is my child mostly introverted or extroverted?
- Does my child like to have a few friends or many friends?
- What group activities in my area are available for homeschoolers?
- Does my child enjoy being alone?
- Does my child get enough sleep? If not, how does this affect his or her learning and relationships?
- Does being comfortable help my child to learn better?
- What is my child's biological clock? What are the best learning times for my child?
- Does my child function far ahead of age peers in one or more areas?
- Is my child struggling to keep up with one or more subjects in ways that affect self-concept and social growth?

- Do I know any adults who could serve as friends or mentors for my child?

Do You Have a Financial Homeschooling Plan?

How will the decision to homeschool affect your budget? Homeschooling can be less expensive than classroom education, especially if children would otherwise go to private schools. For many families, however, homeschooling means a reduction of income and increased expenses for books and other resources.

Before homeschooling, it's good to have an idea of what you are able and willing to spend on home education costs. Remember that many costs—such as swimming lessons (which now count toward physical education) or music lessons—are not really extra costs for homeschooling; they are costs you would incur anyway. If you and your partner talk candidly about the finances of homeschooling before you begin, you'll be less likely to have costs sneak up on you at the end of the year.

Is one parent putting a career or job on hold in order to homeschool? If so, it might be a good idea to keep up with colleagues and work contacts, at least on a social basis. This will help to give your life balance and allow you to stay connected with the adult part of yourself. You can also consider donating your skills to homeschool support groups by sharing what you know with other homeschoolers.

Look into Local Homeschool Support Groups

The appendix lists several contacts for local support groups in each state. Support groups sometimes have specific days when prospective homeschoolers can visit, or you might be able to drop in at any meeting (ask ahead of time). At the very least you should be able to get the name of someone to talk to about the homeschooling experience and how the support group is run. Support groups often distribute newsletters, have Web sites, or manage e-lists and discussion boards. You might be able to subscribe to the newsletter or join an e-list as a prospective homeschooler.

Homeschool support and advocacy groups range from informal groups of parents who get together mainly for conversation and playtime to well-organized institutions with board members and many structured activities. Some groups state clearly that their focus is faith-based homeschooling or secular homeschooling; with other groups you'll need to ask to learn more. Some groups are mainly for young children or teens or homeschoolers with special needs. If you are turned off by the first support group you visit or if their values or reasons for homeschooling differ greatly from yours, look around to see if another group would meet your needs better. You can even start your own group with a few like-minded families.

And remember that not every homeschooling family is part of a support group. We didn't join one for the first six months of our homeschooling, because I wanted to make sure I knew how we wanted to homeschool before I was overly influenced by anyone else. When we did look for a group, I knew exactly what kind of support we needed and was able to find the perfect fit.

Consider Attending a Homeschool Conference

Some state and regional support and advocacy groups host annual homeschool conferences. These conferences are generally held in the spring and summer months, although they can be at any time of the year. Statewide conferences are often scheduled over a weekend and attract hundreds of participants. They offer sessions and workshops for new and veteran homeschoolers, include a book or curriculum fair, and offer plenty of venues for homeschoolers to get together to share ideas. Some may bring in keynote speakers, while others rely on parents to give talks and workshops.

Homeschool conferences can have a very specific focus, such as a Christian home educators' conference or an unschooling conference. Many conferences are inclusive and try to provide sessions for a variety of interests. If you see the word "inclusive" it usually means that both secular and faith-based homeschoolers attend. In Wisconsin, for example, the Wisconsin Parents Association (WPA) holds an inclusive home education conference every spring that attracts a wide variety of homeschoolers of many faiths and with many educational approaches. Also in the spring, the Wisconsin Christian Home Educators Association holds an annual conference and the Greater Milwaukee Catholic Home Educators hosts a Catholic Homeschool Conference and Vendor Fair. Prospective homeschoolers can look for a conference that fits their family.

By attending a homeschool conference, you can learn more about the different ways to learn at home, browse at the book fair, and meet other homeschooling families. In sessions I lead at the WPA conference, I am always impressed by parents who say they are not yet homeschooling, but are trying to learn more about it. Regardless of what they eventually decide, I know they have the potential to be terrific homeschoolers, because they already have shown they care about their children's education, and they are willing to reach out and learn.

How Long Do You Plan to Homeschool?

A very few parents know from the moment their children are born, if not before, that they want to homeschool and they will probably homeschool all the way through high school. Then there are the rest of us. Homeschooling takes us by surprise, and, before we know it, the years add up, and somehow we are veteran homeschoolers.

If you know you want to try homeschooling but are unsure if you can homeschool "forever," rest assured that homeschooling is not a permanent contract. I know families who homeschooled for one year, three years, even one semester, and came away feeling their experience was a success and that it gave them what they needed at the time. Knowing that you can always reassess your decision at the end of the term or year may take some pressure off you or your child.

Some families plan to homeschool for several years, but then stop for some reason. Perhaps a parent needs to return to full-time work or a family crisis makes it hard for a parent to be at home. Sometimes children decide to go to school (and sometimes they decide to return to homeschooling). Any family who has homeschooled even a short length of time learns valuable lessons that all family members carry with them wherever they learn.

Tanya describes her choice to try classroom education, and then her decision to return home:

> When I was in preschool and my brother was about to enter kindergarten, my mother decided to homeschool because the only kindergarten class was thirty-five kids to one adult, it was easier to homeschool two kids than drive one of us thirty minutes to preschool and then take the other one home. We enjoyed homeschooling so much that we stuck with it. The year I was to turn eight, I decided that I wanted to see what school was like, so I went to third grade. I liked school, but at the end of that school year I decided that I liked homeschooling better. This was not because I had trouble in school, on the contrary I was at the head of my class, but I just liked homeschooling better. I wanted to spend time with my homeschooling friends.

Finally, and Most Important, Answer This Question Honestly: Do You Want to Homeschool?

Susan, a homeschool coordinator who has talked to and advised many new homeschooling families, says, "In my opinion, *desire* is far and away the biggest factor in deciding whether to home-school. With desire, just about any obstacle can be overcome. If someone does not really want to homeschool, then I would not recommend it."

Do you want to homeschool? If the answer is no, especially if your children also do not want to homeschool, then you will proba-bly have a difficult time making homeschooling a pleasant experi-ence. If this is the case, you probably haven't made it this far in the book, because you've already seen that homeschooling is not what you want to do.

Do *you* want to homeschool? Or does someone else want you to homeschool? The decision should be yours. If you are willing to give it a try but are unsure, be clear about this rather than pretend you want something you don't.

Do you *want* to homeschool? Not "Do you think you can?" or "Do you feel ready?" or "Do you think you should?" Do you *want* to? If your answer is yes, and if you are a capable and competent par-ent, then you almost certainly *can* homeschool. Take a deep breath, and let's begin!

GETTING STARTED:
PATIENCE, PRACTICE, AND PERSISTENCE

Congratulations! You've made the decision. You are ready to home-school.

Before you buy curricula, join a homeschool association, or draw up a lesson plan, read this section to learn the real secrets of successful homeschooling. Parents who say that homeschooling works well for their children have learned from others or discovered on their own several secrets that make home education more enjoy-able and successful for everyone. Those secrets have little to do with educational standards or expensive resources. They have more to do

with what might be called habits of mind. In talking with dozens of
homeschool parents, I hear three themes over and over: patience,
practice, and persistence:

- Patience not only with our children but with ourselves.
- Practice in the arts of homeschooling and parenting.
- Persistence in the face of problems.

My aim... is to persuade [adults] to look at children, patiently,
repeatedly, respectfully, and to hold off making theories and judg-
ments about them until they have in their minds what most of
them do not now have—a reasonably accurate model of what
children are like.   —John Holt, *How Children Learn*

Until we know who our children are, we cannot create an education that fits their needs. Watching and listening patiently thus
becomes one of the most important and yet most overlooked techniques of homeschooling, as this parent has found:

*I have learned a lot about myself since starting homeschooling, [in-
cluding] that I am not patient and how to be more patient. I have
learned to ask questions before jumping in to solve problems. I have
learned to listen with my heart as well as my ears. I have learned to
make sure I understand the girls before jumping in with suggestions.
And I've learned to have more fun with learning and life.   —Jean*

Patience involves having reasonable expectations for our children and not pressuring them to learn a skill or body of knowledge
before they are ready. Sometimes children tune into a parent's un-
spoken high expectations and respond by "tuning out" or refusing to
work. "He won't do what I tell him!" is a frequent complaint. Perhaps the child is using noncompliance as a way to take some control
over his learning.

Jeffrey Freed, author of *Right-Brained Children in a Left-Brained World,* writes that parents who homeschool can avoid such noncompliance by approaching all learning tasks in a nonjudgmental way.[2] This does not mean that we shouldn't care what our children are learning or take seriously our role as home educator. It does mean, however, that by acting as a facilitator rather than teacher we may have more success, especially with strong-willed children:

> *As I gradually gave up the role of being "teacher" and became a learner, I found joy in a different sense from before. I began learning right along with my children in whatever we were doing. . . . I have found that the more we work together as learners, the more the children are willing to follow a lead I might suggest, which was not the case at first when I was trying to force a direction or an activity.* —Karen

For example, if a homeschooled child balks at writing assignments, the parent can try writing *with* the child rather than simply assigning writing. Perhaps you both can have a time in the morning when you write in your respective journals, side by side. Or you can write a story together, each of you taking a turn with the plot. Some homeschool parents write letters to their children on a regular basis as a way to express affection or to indicate learning assignments. In these ways, the child sees writing as an activity that is important to you and not just as a homeschool task. Writing is presented as a lifelong activity, not a lesson reserved for school-age children.

These kinds of shared and cooperative learning activities also take pressure off the parent to be the perfect teacher. Homeschool parents must be patient with themselves and not try to be Homeschool Super Mom or Dad, as these parents advise:

> *The most important advice I can give for the first year of homeschooling is to take things slowly and not try to cover everything. There is so much adjustment in the first year. . . . You need to adjust to everyone's*

being home all the time, and you need to learn how to get along, how to cope with each other. It's not a good time to try to "do it all" academically or socially. Take it slowly, and learn to enjoy your family again. —Kevin

It has helped me to designate certain times of day as times I am "off duty." Before 8 A.M., if children are awake, they know they need to occupy themselves quietly in their rooms. This gives me time for classical piano, my personal creative outlet. Then, during the little ones' afternoon nap, the older children do silent sustained reading, assuring all of us a quiet break. —Brenda

Children who are physically very active or who are kinesthetic learners call on a different kind of patience, as this homeschool parent has found:

My son has to be personally (and physically) involved in whatever he is studying. He has to draw, write, touch, arrange, flip through pages himself. He loves interrupting and offering his ideas and thoughts. He hates it when he has to sit down calmly and listen to a lecture. —Becca

Homeschool parents of active children can look for ways to integrate activity and learning whenever possible. One homeschool parent allows her son to sit on a big bouncy ball while watching educational videos. Another encourages her daughter to take frequent breaks from sedentary activities to play outside. When our son was younger, he often opened a book on the floor and read it while sitting on his knees and stretching a handful of rubber bands. He also has enjoyed acting out scenes from books or history, either on his own or with friends. These activities not only release psychomotor excitability—preventing an explosion of pent-up tension later in the day—they also promote sensory integration.

Recently I visited a homeschooling family with three very intense, very bright children, but the youngest child, age four, stood out as needing the majority of the parents' patience. As we adults had a conversation in the kitchen, he climbed from chair to chair, sometimes standing on an empty chair to offer a pronouncement before climbing down again, then leaving the room to retrieve some treasure and bringing it back to show me. I don't think he was still for two seconds out of twenty minutes. He was extremely articulate and eager to be a part of the discussion.

What impressed me, though, was both parents' seemingly endless well of patience. They did not try to make him sit still (it would have been fruitless), and they did not apologize for his behavior. He was doing nothing wrong. He was not rude, and he mostly waited his turn before speaking. He was being a very normal, active, highly gifted young child. I couldn't help wondering how much less pleasant the visit would have been had the parents been annoyed with the boy's need to move and allowed themselves to focus on his activity rather than our conversation.

Of course, some of us are born more patient than others. What if we're not naturally patient? What if constant activity or noise or talking drives us crazy?

Patience can be learned and increased with practice. If we lose our patience in the morning, we can refuse to become too discouraged and simply try harder in the afternoon. We can remember that being more patient is a long-term goal that will not be shattered in a moment's weakness. This practice will make us not only better homeschoolers, but better parents.

And when we practice patience with our children, ourselves, and our homeschooling experience, it is easier to be persistent. Persistence—searching for new answers and refusing to give up—is crucial in homeschooling because children are complex. They often

defy labels and expectations of what we think they should be interested in or ready to learn.

To accommodate children's needs, parents must experiment
through trial and error to find what works. Consider the challenge of
a child who claims he isn't interested in history. Rather than trying to
force the child to read textbooks or chide him about the importance
of history, parents can brainstorm about other ways he might come to
enjoy history as well as learn it. Educational history documentaries
often present challenging and interesting concepts that might engage
the child. Humorous history, such as the History News series by Can-
dlewick Press, is enjoyed by many children. Geographical history and
maps, living history museums, historical fiction, and historical crafts
and costumes may also serve as enjoyable ways to learn about history.

Keep in mind that sometimes children will change their learning-style preference or go through growth spurts or sputters, so don't
allow yourself to get locked into a single curriculum or approach. A
good practice is to make available several different approaches to a
subject and allow the child to choose what will work best. This diversity of resources need not be expensive. For example, you might
have available for the study of fractions some colorful and inexpensive workbooks, one or two textbooks, picture books about fractions
from the library, dice (use them to make up division problems),
recipes to be multiplied or divided, and fraction rods (you can make
these yourself with inexpensive wood rods, a jigsaw, and paint).
Watch your child to see what she prefers and whether the preference
changes from day to day or is consistent.

It's true that homeschooling is not the right educational choice
for every family. But when homeschooling is difficult, don't automatically assume you've made the wrong decision. Parents can practice patience and persistence before assuming that home education
won't work for them.

MORE ADVICE ON GETTING STARTED

As a local contact for a state homeschooling organization, I really appreciate the parents who research homeschooling and "do their homework," so to speak. These parents find local sources, they find out the legal ways to homeschool in California, they find out the various philosophies of home education, they join e-mail lists, they check out curriculum, and they equip themselves before beginning the journey. I encourage parents to do this. I also encourage parents to be flexible, to consider their strengths and their children's strengths, and to remember it is a journey that can change quite a bit as you go.... Lastly, I encourage parents to talk with experienced homeschoolers and ask lots of questions. The Internet is rich with homeschooling resources, as well.   —Becca

Advice for New Homeschoolers: You can always try again tomorrow. Children are always learning whether you are actively teaching them or not. It takes time to find a rhythm that fits you and your children. Your rhythm will come. (It won't always keep perfect time though.) It can be a bumpy ride because we're only human. Hang out with other homeschool families when you can. Cut yourself some slack and cut your kids some slack, too. Remember how lucky you are 'cause you get to look at your kids' beautiful faces.   —Debbie

My advice, which I don't always follow, is to relax and enjoy learning together. Get outside and experience the sunshine and rain, the lakes and fields and running down hills. We have a wonderful world to explore.   —Susan

Make your own way for your family and the individuals in it. Then, find like-minded and supportive families. Avoid the negative. Stay open-minded and open to learning more about how children learn, grow, and develop. Keep channels of communication open and well used between your spouse and yourself, your kids and yourself, and with other family members and friends whose support is important to you. Have a goal or "big picture" that you can use to steady yourself during the moments (or days or weeks) when chaos seems to reign.   —*Darla*

9

HOME EDUCATION LAWS

Whether you are thinking about homeschooling or are already homeschooling, it is important to have a thorough understanding of your state's homeschool laws. Likewise, if you are thinking of moving, you will want to research the laws of any states that might be your new home. Suppose you are thinking of homeschooling because you have a friend in another state who has had a successful homeschool experience. However, one of the reasons you are interested in home education is that you want to limit the standardized testing your children receive, and you know that your friend must submit standardized test scores for her children every year. If you check your state's home-education requirements, you may find that you do not need to do any testing. Or, you might find that you need to follow a different attendance schedule, or that some other issue is significantly different.

Most state laws address such issues as: notification of intent to homeschool, ages for which compulsory attendance apply, qualifications of parents, attendance requirements, whether the family is considered a private school, subjects and curriculum to be covered, extent of contact families may have with local school districts, testing and assessment, record keeping, and education for children's special needs. To get a better understanding of homeschool requirements, we can look at each of these issues in detail with some examples.

The following information is a general overview of what to look for when researching home-education regulations. It is intended neither as legal advice nor as a complete guide to state laws, and is based on information current as of this book's publication. For more specific details, please refer to your state's statutes and get in touch with a state or local homeschooling group.

NOTIFICATION

Families new to homeschooling need to know if they must notify their state department of education or their local school authorities of their intent to homeschool. In some states, no notification is required. In other states, all homeschoolers must submit notification by a certain date (usually summer or fall), and some states require only those children who are leaving school—rather than those never having gone to school—to fill out notification forms. Illinois, for example, requires no notification. Ohio requires families to provide written notification to the local school superintendent each year before beginning homeschooling. Oregon requires notification when a child starts homeschooling or moves to another district. Homeschool parents can take care to follow whatever notification procedures their state requires without providing more information than is necessary.

AGE OF COMPULSORY ATTENDANCE

All states set ages when children must begin schooling and when they are no longer bound by compulsory attendance. In Wisconsin, for example, all children ages six (by September 1) through eighteen fall under compulsory education laws. In California, the ages are also six and eighteen; however the birth date cutoff is December 2. In Florida, compulsory school ages are six (by February 1) to sixteen; in Idaho, seven to sixteen. These ages almost always are the same as for students attending public or private schools.

PARENT QUALIFICATIONS

Some states require no qualifications other than being a child's parent or legal guardian. Parents who are considering homeschooling already are showing care, commitment, and a desire to give their children a good education.[1] Parenting involves other skills important to home education: a willingness to learn from mistakes and the ability to find and use information. Whether homeschooling is the right decision for a particular family is a separate issue; however, almost anyone with a desire to homeschool is qualified to do so.

Some states require homeschool parents to have a specific level of formal education or to work with certified teachers to fulfill the law. In the state of Washington, for example, parents can qualify to homeschool their children in one of four ways: have forty-five units of college-level credit; attend a qualifying course for parents; have a certified teacher meet with the homeschooled child for one hour per week; or be deemed "sufficiently qualified to provide home-based instruction" by the local school district's superintendent. In North Dakota, if parents possess a high school diploma or a GED but not a teaching certificate or a baccalaureate degree, they must either prove that they meet or exceed minimum qualifying scores on the national teacher exam, or be monitored by a certified teacher for the first two years. Monitoring continues after two years if a child scores lower than the fiftieth percentile on standardized tests.

Parental requirements for homeschooling raise interesting questions and issues. Private schools often do not have to hire certified teachers, even though these teachers are working with more children at once than homeschool parents. Much of the teacher training required for certification isn't useful for home education, where keeping a group of students on task and developing lesson plans for many children of a similar age are not applicable. For one-on-one learning, the most important qualifications are a willingness to find and do whatever works best and a deep knowledge of the learner. In these qualifications, parents are the experts, regardless of their degrees or certifications.

ATTENDANCE

Attendance in some kind of education program is compulsory for all children in the United States. Some states specify a number of hours, days, or months required of homeschooled children. Not all states require that attendance records be kept or submitted (see Record Keeping, below). Mississippi homeschool parents set their own attendance requirements as necessary to progress from grade to grade. In South Dakota, homeschoolers are required to attend for a length of time equivalent to that of public school students. California homeschoolers can choose the option of using a private tutor for a minimum of 175 days per year, three hours per day. Wisconsin homeschool parents must provide "at least 875 hours of instruction each year." In Montana, the hours of required attendance are different for younger and older children: 720 hours per year for grades one through three and 1,080 hours per year for grades four through twelve. Homeschoolers in the District of Columbia are required to follow the periods when public schools are in session.

Depending on the state of residence, parents may have much flexibility with their home education calendar. Unless the law restricts attendance to traditional school days or months, families can divide the hours or days throughout the year, homeschool during the summer and take time off for travel in the winter, or homeschool late afternoons and evenings rather than in the morning. In Wisconsin, for example, if a child wants to do a little homeschooling every day of the year, including weekends, two and one-half hours per day would meet the requirement of 875 hours annually.

PRIVATE SCHOOL STATUS

In some states, homeschools are considered to be private schools, with homeschooling requirements falling under private school regulations. The paperwork may be the same as for private schools or it may be specific to homeschoolers. In California, for example, where homeschoolers have four options for homeschooling legally, parents can qualify their homeschooling as a private school by meeting private school requirements and filing a private school affidavit.

Alaska and Louisiana are examples of states that offer the option of parents setting up and operating a homeschool or qualifying as a private school. Other states, like Arkansas and Vermont, require all families who learn at home to set up and operate a homeschool. The difference is mainly one of legal definition. It's important to know what your home education is classified as, because this may affect the services available to you and whether private school legislation applies to your homeschooling.

SUBJECTS AND CURRICULUM

States may or may not list specific subjects for homeschoolers to study. Many states require that homeschools offer the same curriculum as public schools in the state, while a few have no subject requirements. Kentucky's required homeschool subjects are reading, writing, spelling, grammar, history, mathematics, and civics. Penn-sylvania homeschoolers who choose the option of a private tutor must provide their high school children with instruction in English language, literature, speech and composition, science, biology, chemistry, geography, social studies, civics, economics, world, U.S., and Pennsylvania history, a foreign language, general mathematics and statistics, algebra and geometry, art, music, physical education, health and physiology, and safety and fire prevention.

Again, it is important to know what is and is not required in the state where you live. In Wisconsin, for example, parents are to provide "a sequentially progressive curriculum of fundamental instruction" in reading, language arts, math, social studies, science, and health. There are no requirements for how much time is to be spent on each subject, other than the number of total hours per year. The proscribed "sequentially progressive curriculum" simply means that children are progressing in their understanding by building on existing knowledge.[2] Parents may choose to fulfill this requirement with a traditional, grade-based curriculum or by letting children follow their own curiosity to the next step in learning. By understanding what is required as well as what is allowed by law, families give themselves as many choices as possible.

CONTACT WITH LOCAL SCHOOL DISTRICT

Depending on where they live, homeschool families may be required to have some contact with their local school district personnel. This could be when they notify them of their intent to homeschool, when they submit test scores or other paperwork, or at quarterly or annual reviews. In Hawaii, for example, homeschoolers file a notice of intent with the principal of the public school that the child would otherwise attend. In North Dakota, standardized test scores for homeschoolers are filed with the local superintendent. In other states, local education officials have no part in the fulfillment of homeschool requirements, and home-education parents deal only with the state government.

TESTING AND ASSESSMENT

One thing that all parents thinking of homeschooling will want to know is if their state requires them to submit standardized-test scores. Several states require no testing of homeschooled children. Other states call for homeschooled children to take whatever standardized tests are given to public school students, or to be given some other state-approved standardized test at regular intervals. In some cases, parents have the option of waiving the standardized-test requirement if they agree to have their child's progress reviewed by a certified teacher, principal, or superintendent, or an advisory board set up for this purpose.

Parents will also want to know the legal effect of any standardized-test scores submitted to the state. As we saw previously, North Dakota parents who have no education past a high school diploma and do not have a teaching certificate must be monitored by a certified teacher for as long as the children's standardized-test scores fall below the fiftieth percentile. In New York, homeschools in which children who score below the thirty-third percentile on standardized tests can be put on probation.

Who is accountable for a child's education, and what rights do parents have in their children's education? These questions are at the heart of the issue of mandated testing, and the answers have many repercussions. If the parents are ultimately responsible and accountable for knowing if their children are getting a good education, then they would have the choice to test or not as part of that accountability. Several states allow parents to waive standardized testing for their children in public schools. However, the issue becomes complicated.

The *Wall Street Journal* has reported that many of the children who opt out of these tests would score well on them. This affects schools' overall test scores, reputations, and, sometimes, eligibility for government funding, which in turn causes the schools to pressure the students to be tested. To convince students not to waive the tests, some states even offer scholarships of up to $2,500 to students who score well.[3] If parents do need to have their children tested to fulfill legal homeschool requirements, they can remember that test scores do not define a child's intelligence or even his knowledge base, and sometimes have more to do with schools' reputations and funding than with measuring an individual student's progress or potential.

RECORD KEEPING

Records that homeschool parents may be required to keep include attendance logs, immunization records, quarterly progress reports, summaries of curricula and textbooks used, reading logs, test results, or samples of writing or other work. In some states, no records are required. In others, record keeping is a large part of what it means to homeschool legally.

Attendance records may need to show hours or days of study. Parents who live in a state that requires a certain number of education hours per calendar year can determine, on average, how many hours per day are necessary to meet the requirement, then use a calendar to keep track of any day when that minimum is met. Other families keep more detailed attendance logs that include what was studied, field trips taken, or hours per subject. What works best will differ from family to family. Several books and Internet sites offer forms that homeschoolers can use for record keeping, or states may supply specific forms.

Families who live in states that require extensive record keeping can look for support from other homeschoolers in fulfilling the law without doing more than is necessary. For example, in South Carolina, where homeschools are approved by local boards of trustees, several support groups are available throughout the state to help parents learn how to keep the necessary portfolios, academic records, and progress reports. One group even offers several free forms online to make the process easier. Parents who live in states with little or no required record keeping can think about what kinds of records they would want to keep for themselves. The Wisconsin Parents Association suggests that parents think not only about the records that are required by law, but also any personal records that the family finds important or might need at a later date.[4] Especially when homeschooled children start thinking about college, internships, or jobs, they will want to have records that they can use for portfolios, applications, and transcripts.

SPECIAL NEEDS

A few states specify homeschool guidelines for children with special learning needs, such as autism or attention deficit hyperactivity disorder. Parents who have children who have special needs or who struggle to learn should check if their state has specific guidelines for them, or if homeschooled children are eligible to receive any special-needs services. In North Dakota, for example, homeschool parents of developmentally disabled children select an individualized education program team, which prepares progress reports three times annually. In Iowa, parents of children identified with special needs must gain the approval of the local special-education director.

Several support groups and Internet groups are available both nationwide and in local areas for families who homeschool children with Asperger's Syndrome, learning differences, and other special needs. Many parents appreciate the fact that they can meet their children's extreme needs without calling undue attention to them or unnecessarily labeling them. In homeschooling, each child's needs and strengths are singled out and nurtured, regardless of where they might fall on a bell curve.

If all the legal requirements make homeschooling sound complicated, keep in mind that each state's laws are usually easy to understand and follow. We can remember that the essence of home education—which is separate from the legal requirements—is simply children learning under the guidance of their families.

EPILOGUE

HOMESCHOOLING FOR THE TWENTY-FIRST CENTURY

The traditional organization of schooling is intellectually and morally inadequate for contemporary society. We live in an age troubled by social problems that force us to reconsider what we do in schools. Too many of us think that we can improve education merely by designing a better curriculum, finding and implementing a better form of instruction, or instituting a better form of classroom management. These things won't work.

> — *Nel Noddings, "A Morally Defensible*
> *Mission for Schools in the 21st Century"*

We have just begun to see the emergence of home schooling as an important national phenomenon. Unless the needs of parents are met in different ways, it is likely that home schooling will have a large impact on the school as an institution in coming decades.

> — *Kurt Bauman, U.S. Census Bureau*

Parents have a prior right to choose the kind of education that shall be given to their children.

> — *Universal Declaration of Human Rights*

As with so much of homeschooling, writing this book has been an education for me. I began by wanting to show readers that the homeschooling option works. Not until I put the words on paper did I realize exactly why it works, not just for our family, but for all the families who choose to come home each fall.

One of the parents I interviewed said that we shouldn't worry too much about whether our homeschooled kids have missed anything academically, because they probably have, and when they need to learn it, they will. It's taken a few years of my own home education to understand what she means and to see home education in its larger context. We've all read about the academic success of homeschoolers, the studies and statistics that show they can test as well as their classroom peers and get into the same colleges. We also read about homeschoolers who are socialized, able to get along with others, and follow the rules. While these articles have helped to legitimize homeschooling and address some valid concerns, they miss the point.

Academic success and socialization are not why homeschooling works.

Homeschooling works because it offers a unique and exciting vision of how education, family life, social needs, and the individual can come together at a time when we are questioning the kind of life we want for our children in this new century. Homeschooling gives families the time and space to live together for more than a few minutes each day. Homeschooled children learn that they don't need to wait until they are eighteen or twenty-two to take their place in society. Home education is tailored to all the child's needs—intellectual, emotional, social, psychological, and spiritual.

Families may homeschool initially to solve a school problem or to improve academic achievement, but those who continue to homeschool year after year do it for one reason: they like being a homeschooling family.

Homeschoolers don't need to ask or prove how they can make their homes more like schools. Our homes are—and always have been—places of learning. The question for the twenty-first century, for those children whose parents do not homeschool, is whether we can make schools more like homeschooling.

APPENDIX

CONTACT INFORMATION FOR STATE LAWS AND SUPPORT GROUPS

Use the following links to learn more about homeschooling legislation in your state and to get in touch with support groups and associations. Homeschool groups come in many shapes and sizes: informal groups of families that meet for socializing, formal homeschool co-ops that offer classes for fees, advocacy associations that lobby on behalf of homeschoolers, and clearinghouses for information about other groups and about homeschooling.

When possible, links are provided to state government pages for the text of homeschool laws and regulations. When there is no convenient link to the state government site (or for states where homeschool laws are not collected in one place), I have provided an alternative link for the information, either from the *Home Education Magazine* site or a support group.

Knowing the law is each homeschooling family's responsibility. The information here and throughout this book is informational only and meant to help you to decide if homeschooling is right for you. It is not intended as legal advice. For more detailed information, contact one of the support groups or associations listed below. When the name of the state organization does not clearly indicate whether the group has a specific focus or geographical area, I have provided that information in parentheses. Please keep in mind that this does not necessarily imply that the group does not have an open membership policy.

NATIONAL HOMESCHOOL ASSOCIATIONS

American Homeschool Association
http://www.americanhomeschoolassociation.org/

Home School Legal Defense Association (HSLDA)
http://www.hslda.org/

NATIONWIDE RESOURCES

These sites offer further links to state legal information and support groups.

Home Education Magazine (HEM)
http://homeedmag.com/

A to Z Home's Cool Homeschooling Web Site
http://homeschooling.gomilpitas.com/

Leaping from the Box
http://www.leapingfromthebox.com/

Learn in Freedom!
http://www.learninfreedom.org/

ALABAMA

Legal Regulations

Home Education Magazine: Alabama Laws and Regulations
http://www.homeedmag.com/lawregs/alabama.html

Selected Support and Advocacy Groups

Teaching at Home (central Alabama)
http://homeschoolblogger.com/teachingathome

Unschooling in Alabama
http://www.pipeline.com/~wdkmg/homeschool/unschool.htm

Christian Home Education Fellowship of Alabama
http://www.chefofalabama.org/

ALASKA

Legal Regulations

Home Education Magazine: Alaska Laws and Regulations
http://www.homeedmag.com/lawregs/alaska.html

Selected Support and Advocacy Groups

Alaska Homeschool Network (Yahoo group)
http://groups.yahoo.com/group/AlaskaHomeschoolNetwork/

Alaska Private and Home Educators Association (Christian)
http://www.aphea.org/

ARIZONA

Legal Regulations

State Legislature: Revised Statutes on Homeschooling
http://www.azleg.state.az.us/ars/15/00802.htm

Selected Support and Advocacy Groups

Home Education Network of Arizona
http://www.hena.us/

Arizona Families for Home Education (open membership, Christian board)
http://www.afhe.org/

Phoenix Unschoolers (Yahoo group)
http://groups.yahoo.com/group/Phoenix_Unschoolers/

ARKANSAS

Legal Regulations

Department of Education: Home School Office
http://arkedu.state.ar.us/schools/schools_home.html

Selected Support and Advocacy Groups

Live and Learn (central Arkansas)
http://www.geocities.com/live-and-learn.geo/

Christian Home Educators of Northeast Arkansas
http://www.neache.org/

CALIFORNIA

Legal Regulations

Home Education Magazine: California Laws and Regulations
http://www.homeedmag.com/lawregs/california.html

Selected Support and Advocacy Groups

California Homeschool Network
http://californiahomeschool.net/default.htm

Homeschool Association of California
http://www.hsc.org/

Christian Home Educators Association of California
http://www.cheaofca.org/

COLORADO

Legal Regulations

Department of Education: Home Schooling in Colorado
http://www.cde.state.co.us/cdeedserv/homeschool.htm

Selected Support and Advocacy Groups

Rocky Mountain Education Connection
http://rmec-online.com/index.html

Mile High Homeschoolers (Denver)
http://www.milehighhomeschoolers.com/

Christian Home Educators of Colorado
http://www.chec.org/

CONNECTICUT

Legal Regulations

Home Education Magazine: Connecticut Laws and Regulations
http://www.homeedmag.com/lawregs/connecticut.html

Selected Support and Advocacy Groups

Connecticut Homeschool Network
http://www.cthomeschoolnetwork.org/

The Education Association of Christian Homeschoolers
http://www.teachct.org/

DELAWARE

Legal Regulations

Home Education Magazine: Delaware Laws and Regulations
http://www.homeedmag.com/lawregs/delaware.html

Selected Support and Advocacy Groups

Delaware Home Education Association
http://www.dheaonline.org/

Tri-State Home School Network (Christian)
http://www.tristatehomeschoolnetwork.org/

FLORIDA

Legal Regulations

Home Education Magazine: Florida Laws and Regulations
http://www.homeedmag.com/lawregs/florida.html

Selected Support and Advocacy Groups

Florida Parent-Educators Association
http://www.fpea.com/

Mid-Florida Homeschoolers (central Florida)
http://www.midflhomeschoolers.com/

Splash! (southern Florida)
http://www.splashfl.org/

GEORGIA

Legal Regulations

HEIR of Georgia: Home Study Law (pdf file)
http://www.heir.org/pdf-files/homestudylaw.pdf

Selected Support and Advocacy Groups

Georgia Home Education Association (Christian)
http://www.ghea.org/

Home Education Information Resource of Georgia
http://www.heir.org/

HAWAII

Legal Regulations

Department of Education: Home Schooling
http://doe.k12.hi.us/myschool/homeschool/

Selected Support and Advocacy Groups

Hawai'i Homeschool Association (Oahu)
http://www.hawaiihomeschoolassociation.org/

Christian Homeschoolers of Hawaii
http://www.christianhomeschoolersofhawaii.org/

IDAHO

Legal Regulations

Department of Education: Home School
http://www.sde.idaho.gov/HomeSchool/default.asp

Selected Support and Advocacy Groups

Southeast Idaho Homeschool Association (Pocatello area)
http://www.eyedocgreg.com/homeschool/index.htm

Christian Homeschoolers of Idaho State
http://www.chois.org/

ILLINOIS

Legal Regulations

Illinois H.O.U.S.E.: Summary of Illinois Laws that Pertain to Homeschooling
http://www.illinoishouse.org/a14.htm

Selected Support and Advocacy Groups

Illinois H.O.U.S.E.
http://www.illinoishouse.org/

Northside Unschoolers of Chicago
http://www.northsideunschoolers.org/

Illinois Christian Home Educators
http://www.iche.org/pages/law.php

INDIANA

Legal Regulations

Department of Education: Private Education
http://www.doe.state.in.us/sservices/hse.htm

Selected Support and Advocacy Groups

Indiana Home Educators' Network
http://www.ihen.org/default_portal.htm

Indiana Foundation for Home Schooling
http://www.ifhsonline.org/

Indiana Association of Home Educators (Christian)
http://www.inhomeeducators.org/

IOWA

Legal Regulations

State Legislature: Private Instruction Regulations
http://www.legis.state.ia.us/IACODE/2003/299A/1.html
http://www.legis.state.ia.us/IACODE/2003/299A/3.html
http://www.legis.state.ia.us/IACODE/2003/299A/4.html

Selected Support and Advocacy Groups

Iowa Home Educators
http://www.iahomeeducators.org/

Network of Iowa Christian Home Educators
http://www.the-niche.org/

KANSAS

Legal Regulations

Department of Education: Nonaccredited Private Schools
http://www.ksde.org/Default.aspx?tabid=1684

Selected Support and Advocacy Groups

EDUCATE (secular, Johnson County)
http://www.educateks.com/

LEARN (secular, Kansas City)
http://www.kclearn.org/

Christian Home Educators Confederation of Kansas
http://www.kansashomeschool.org/

KENTUCKY

Legal Regulations

Department of Education: Home Schooling
http://www.education.ky.gov/KDE/About+Schools+and+Districts/Home+Schooli
  ng+in+Kentucky/

Selected Support and Advocacy Groups

Kentucky Home Education Association
http://www.khea.8k.com/

Christian Home Educators of Kentucky
http://www.chek.org/

LOUISIANA

Legal Regulations

Department of Education: Registered Nonpublic Schools
http://www.doe.state.la.us/lde/eia/2158.html

Selected Support and Advocacy Groups

Louisiana Home Education Network
http://www.la-home-education.com/

Christian Home Educators Fellowship of Louisiana
http://www.chefofla.org/

MAINE

Legal Regulations

Department of Education: Home Instruction
http://www.maine.gov/education/hs/index.html

Selected Support and Advocacy Groups

Central Maine Self-Learners Homeschool Group
http://members.aol.com/cmslhomeschool/

Homeschoolers of Maine (Christian)
http://homeschoolersofmaine.org/index.htm

MARYLAND

Legal Regulations

Department of Education: Home Instruction
http://www.marylandpublicschools.org/MSDE/nonpublicschools/npdocs/fact_she
ets/np_fact_home_instruction.htm

Selected Support and Advocacy Groups

Maryland Home Education Association
http://www.mhea.com/

Maryland Association of Christian Home Educators
http://www.machemd.org/

MASSACHUSETTS

Legal Regulations

Department of Education: Home Schooling
http://www.marylandpublicschools.org/MSDE/divisions/studentschoolsvcs/studen
    t_services_alt/home_schooling/

Selected Support and Advocacy Groups

Massachusetts Home Learning Association
http://www.mhla.org/

Massachusetts Homeschool Association of Parent Educators (Christian)
http://www.masshope.org/

MICHIGAN

Legal Regulations

Department of Education: Nonpublic and Home Schools
http://www.michigan.gov/mde/0,1607,7–140–6530_6564_35175---,00.html

Selected Support and Advocacy Groups

The Homeschool HUB
http://www.hshub.org/

Information Network for Christian Homes
http://inch.org/

MINNESOTA

Legal Regulations

Home Education Magazine: Minnesota Laws and Regulations
http://www.homeedmag.com/lawregs/minnesota.html

Selected Support and Advocacy Groups

Minnesota Homeschoolers' Alliance
http://www.homeschoolers.org/

Minnesota Association of Christian Home Educators
http://www.mache.org/

MISSISSIPPI

Legal Regulations

Home Education Magazine: Mississippi Laws and Regulations
http://www.homeedmag.com/lawregs/mississippi.html

Selected Support and Advocacy Groups

Parent Educators and Kids (PEAK)
http://www.peaknetwork.org/

Mississippi Home Educators Association (Christian)
http://www.mhea.net/

MISSOURI

Legal Regulations

Home Education Magazine: Missouri Laws and Regulations
http://www.homeedmag.com/lawregs/missouri.html

Selected Support and Advocacy Groups

Families for Home Education
http://www.fhe-mo.org/

Missouri Association of Teaching Christian Homes
http://www.match-inc.org/

MONTANA

Legal Regulations

Office of Public Instruction: Home School Information (pdf file)
http://www.opi.mt.gov/pdf/Measurement/HomeSchoolPacket.pdf

Selected Support and Advocacy Groups

Montana Coalition of Home Educators
http://www.mtche.org/

Bozeman Homeschool Network (Yahoo group)
http://groups.yahoo.com/group/BHN/

NEBRASKA

Legal Regulations

Department of Education: Exempt (Home) School Program
http://ess.nde.state.ne.us/OrgServices/ExemptSchools/Default.htm

Selected Support and Advocacy Groups

Nebraskahomeschool (Yahoo group)
http://groups.yahoo.com/group/Nebraskahomeschool/

Nebraska Christian Home Educators Association
http://www.nchea.org/

NEVADA

Legal Regulations

Department of Education: Homeschooling
http://www.doe.nv.gov/schools/homeschooling/

Selected Support and Advocacy Groups

Nevada Homeschool Network
http://www.nevadahomeschoolnetwork.com/

Home Educators of Faith (Christian)
http://www.homeeducatorsoffaith.org/

NEW HAMPSHIRE

Legal Regulations

Statutes: Home Education
http://www.gencourt.state.nh.us/rsa/html/NHTOC/NHTOC-XV-193-A.htm

Selected Support and Advocacy Groups

New Hampshire Homeschooling Coalition
http://www.nhhomeschooling.org/

Christian Home Educators of New Hampshire
http://www.mv.com/ipusers/chenh/index.php?page=home

NEW JERSEY

Legal Regulations

Department of Education: Homeschooling FAQ
http://www.state.nj.us/education/genfo/overview/faq_homeschool.htm

Selected Support and Advocacy Groups

New Jersey Homeschool Association
http://www.geocities.com/jerseyhome/

Education Network of Christian Homeschoolers of New Jersey
http://www.enochnj.org/

NEW MEXICO

Legal Regulations

Public Education Department: Homeschools
http://nmhomeschools.org/

Selected Support and Advocacy Groups

Albuquerque Home Learners
http://www.abqhomelearners.org/

Christian Association of Parent Educators of New Mexico
http://www.cape-nm.org/

NEW YORK

Legal Regulations

Education Department: Home Instruction
http://www.emsc.nysed.gov/nonpub/homeinstruction.html

Selected Support and Advocacy Groups

New York Home Educators Network
http://www.nyhen.org/

New York City Home Educators Alliance
http://www.nychea.org/

Loving Education at Home (Christian)
http://www.leah.org/

NORTH CAROLINA

Legal Regulations

Division of Non-Public Education: Home School Information
http://www.ncdnpe.org/hhh103.htm

Selected Support and Advocacy Groups

Homeschool Alliance of North Carolina
http://www.ha-nc.org/

Triad Home Educator Support Co-op (Christian, Triad area)
http://groups.yahoo.com/group/THE-SCOOP/

NORTH DAKOTA

Legal Regulations

Department of Public Instruction: Home Education (pdf file)
http://www.legis.nd.gov/cencode/t151c23.pdf

Selected Support and Advocacy Groups

North Dakota Home School Association (Christian)
http://ndhsa.org/

OHIO

Legal Regulations

Home Education Magazine: Ohio Laws and Regulations
http://www.homeedmag.com/lawregs/ohio.html

Selected Support and Advocacy Groups

Ohio Home Educators Network (Northeast Ohio)
http://www.ohiohomeeducators.net/

Christian Home Educators of Ohio
http://www.cheohome.org/

OKLAHOMA

Legal Regulations

Home Education Magazine: Oklahoma Laws and Regulations
http://www.homeedmag.com/lawregs/oklahoma.html

Selected Support and Advocacy Groups

Home Educators Resource Organization of Oklahoma
http://www.oklahomahomeschooling.org/

Christian Home Educators Fellowship of Oklahoma
http://www.chefok.org/

OREGON

Legal Regulations

Department of Education: Home Schooling
http://www.ode.state.or.us/search/results/?id=74

Selected Support and Advocacy Groups

Oregon Home Education Network
http://www.ohen.org/

Oregon Christian Home Education Association Network
http://www.oceanetwork.org/highres.cfm

PENNSYLVANIA

Legal Regulations

Department of Education: Home Education and Private Tutoring
http://www.pde.state.pa.us/home_education/site/default.asp

Selected Support and Advocacy Groups

Pennsylvania Home Educators Association
http://www.phea.net/

Christian Homeschool Association of Pennsylvania
http://www.chapboard.org/

RHODE ISLAND

Legal Regulations

Statutes: Compulsory Attendance and Approval of Private Schools
http://www.rilin.state.ri.us/Statutes/TITLE16/INDEX.HTM

Selected Support and Advocacy Groups

Secular in the Ocean State
http://www.soshomeskoolri.org/

Rhode Island Christian Home Educators
http://richeshomeschool.org/

SOUTH CAROLINA

Legal Regulations

Home Education Magazine: South Carolina Laws and Regulations
http://www.homeedmag.com/lawregs/southcarolina.html

Selected Support and Advocacy Groups

Tri-County Educational Association of Community Homeschoolers (Charleston)
http://www.teach-hs.org/

Carolina Homeschooler (for homeschoolers who use S.C.'s third option)
http://www.carolinahomeschooler.com

Christian Homeschooler's Association of South Eastern South Carolina
http://chasesc.com/

SOUTH DAKOTA

Legal Regulations

Department of Education: Homeschooling/Alternative Instruction
http://doe.sd.gov/oatq/accreditation/altinstruction.asp

Selected Support and Advocacy Groups

South Dakota Home School Association of Sioux Falls
http://www.sdhsa.org/SDHSA/Home.html

South Dakota Christian Home Educators
http://www.sdche.org/

TENNESSEE

Legal Regulations

Department of Education: Requirements for Homeschools
http://www.state.tn.us/education/homeschool/

Selected Support and Advocacy Groups

Spring Hill Area Parent Educators
http://groups.yahoo.com/group/SHAPE_announcements_group/

Home Education Association Rutherford Tennessee (Christian)
http://www.rutherfordheart.com/

TEXAS

Legal Regulations

Texas Education Agency: Home School Information
http://www.tea.state.tx.us/home.school/

Selected Support and Advocacy Groups

Austin Area Homeschoolers
http://www.geocities.com/austinareahomeschoolers/

Houston Unschoolers Group
http://www.geocities.com/Athens/Delphi/1794/hug.html

Home School Texas (Christian)
http://www.homeschooltexas.com/

UTAH

Legal Regulations

Office of Education: Homeschooling
http://www.usoe.k12.ut.us/curr/homeschool/default.htm

Selected Support and Advocacy Groups

Utah Home Education Association
http://www.uhea.org/

Utah Christian Home School Association
http://www.utch.org/

VERMONT

Legal Regulations

Department of Education: Home Study
http://education.vermont.gov/new/html/pgm_homestudy.html

Selected Support and Advocacy Groups

Vermont Home Education Network
http://www.vhen.net/

Vermont Association of Home Educators
http://www.vermonthomeschool.org/

VIRGINIA

Legal Regulations

Department of Education: Home Instruction in Virginia
http://www.pen.k12.va.us/VDOE/Parents/factshee.html

Selected Support and Advocacy Groups

Organization of Virginia Homeschoolers
http://www.vahomeschoolers.org/

Home Educators Association of Virginia (biblical worldview)
http://www.heav.org/

WASHINGTON

Legal Regulations

Superintendent of Public Instruction: Office of Private Education
http://www.k12.wa.us/PrivateEd/HomeBasedEd/

Selected Support and Advocacy Groups

Washington Homeschool Organization
http://www.washhomeschool.org/

Washington Natural Learning Association (unschooling)
http://wnla.tripod.com/id36.html

Christian Homeschool Network of Washington
http://www.chnow.org/

WEST VIRGINIA

Legal Regulations

Home Education Magazine: West Virginia Laws and Regulations
http://www.homeedmag.com/lawregs/westvirginia.html

Selected Support and Advocacy Groups

West Virginia Home Educators Association
http://www.wvhea.org/

Christian Home Educators of West Virginia
http://www.chewv.org/

WISCONSIN

Legal Regulations

Department of Public Instruction Home-Based Private Education Program
http://dpi.wi.gov/sms/homeb.html

Selected Support and Advocacy Groups

Wisconsin Parents Association
http://homeschooling-wpa.org

Wisconsin Christian Home Educators Association
http://www.wisconsinchea.com/

WYOMING

Legal Regulations

Home Education Magazine: Wyoming Law and Regulations
http://www.homeedmag.com/lawregs/wyoming.html

Selected Support and Advocacy Groups

Rocky Mountain Education Connection
http://www.rmec-online.com/index.html

Homeschoolers of Wyoming (Christian)
http://www.homeschoolersofwy.org/

NOTES

INTRODUCTION

1. Brain Ray, "Homeschoolers on to College: What Research Shows Us," *Journal of College Admission* (Fall 2004): 5–11.
2. Michelle Bates Deakin, "Schoolhouse Rocked," *Boston Globe*, March 21, 2004, http://www.boston.com/news/globe/magazine/articles/2004/03/21/schoolhouse_rocked/?page=1.
3. John-John Williams IV, "It's Gym Time at Dance School," *Baltimore Sun*, November 11, 2006.
4. Elizabeth Maker, "Home Schoolers Find Strength in Numbers," *New York Times*, February 11, 2007, sec. 14CN.
5. Cynthia Leonor Garza, "When Home Become School: With a Few Design Tricks, Extra Rooms Convert to Perfect Learning Spaces," *Houston Chronicle*, September 9, 2006, sec. E.
6. Angie Kiesling, "The Quiet 'Boom' Market," *Publisher's Weekly*, July 9, 2001, 23–25.
7. Amy Esbenshade Hebert, "Finance Lessons for Home Schoolers: A Custom-Tailored Classroom Has a Lot of Pluses, but You'll Pay a Price," *Kiplinger's Personal Finance*, March 2007, 88–89.
8. Sue Shellenbarger, "Extreme Juggling: Parents Home-School the Kids While Holding Full-Time Jobs," *Wall Street Journal*, September 14, 2006, sec. D.
9. Leslie Fulbright, "Blacks Take Education into Their Own Hands: Once Dominated by Whites, Homeschooling Appeals to More African Americans," *San Francisco Chronicle*, September 25, 2006, sec. A.
10. Christina Morrow, "A Growing Option," *New Pittsburgh Courier*, September 27–October 3, 2006, sec. A.

CHAPTER ONE

1. Mitchell L. Stevens, "The Normalisation of Homeschooling in the USA," *Evaluation and Research in Education* 17, nos. 2 and 3 (2003): 90–100.

2. Larry and Susan Kaseman, "Does Homeschooling Have to Be Political?" *Home Education Magazine,* January–February 2004, http://www.home-edmag.com/HEM/211/jftch.html.

3. Larry and Susan Kaseman, "Foundations of the Rights and Responsibilities of Homeschooling Parents," *Home Education Magazine,* May–June 1996, http://www.homeedmag.com/INF/FREE/free_fndrf.html.

4. U.S. Department of Education, "School Choices for Parents," http://www.ed.gov/parents/schools/choice/definitions.html#hs.

5. Parker Palmer, *The Courage to Teach* (San Francisco: Jossey-Bass, 1998).

6. Mary Elaine Jacobsen, *The Gifted Adult: A Revolutionary Guide for Liberating Everyday Genius* (New York: Ballentine, 2000).

7. Tim Madigan, *I'm Proud of You: My Friendship with Fred Rogers* (New York: Gotham, 2006).

8. Roy Underhill, "'The Constancy of Change': Remarks at the Opening Session of WPA's 19th Annual Home Education Conference on May 4, 2002," *WPA Newsletter* 72 (2002): 7–8.

CHAPTER TWO

1. John Holt, *Teach Your Own: A Hopeful Path for Education,* revised and updated by Patrick Farenga (New York: Perseus, 2003).

2. David Albert, *And the Skylark Sings with Me: Adventures in Homeschooling and Community-Based Education* (British Columbia, Canada: New Society Publishers, 1999).

3. Kurt Bauman, "Home Schooling in the United States: Trends and Characteristics," U.S. Census Bureau, Population Division (2001) http://www.census.gov/population/www/documentation/twps0053.html; U.S. Department of Education, "1.1 Million Homeschooled Students in the United States in 2003," July 2004, http://nces.ed.gov/pubsearch/pubsinfo.asp?pubid=2004115.

4. Bauman, "Home Schooling in the United States."

5. Wisconsin Parents Association, *Homeschooling in Wisconsin: At Home with Learning,* 6th ed. (Madison, Wis.: Wisconsin Parents Association, 2006).

6. Wisconsin Department of Public Instruction, "Comparison of Enrollment Trends in Home-Based Private Educational Programs," http://dpi.wi.gov/sms/xls/hbtrends.xls.

7. Linda Dobson, *Homeschoolers' Success Stories: 15 Adults and 12 Young People Share the Impact That Homeschooling Has Made on Their Lives* (Roseville, Cal.: Prima Publishing, 2000).

8. John Locke, "Some Thoughts Concerning Education," (1693), http://www.fordham.edu/halsall/mod/1692locke-education.html.

9. National Center for Policy Analysis, *Policy Digest: Education,* March 1996, http://www.ncpa.org/pi/edu/pdedu/pdedu26.html.

10. Edward E. Gordan, *Literacy in America: History Journey and Contemporary Solutions* (Westport, Conn.: Praeger, 2003).

11. Ivan Illich, *Deschooling Society* (London, New York: Marion Boyers, 1971).

12. Milton Friedman, "Public Schools: Make Them Private," Cato Institute Briefing Paper 23, http://www.cato.org/pubs/briefs/bp–023.html.

13. Mitchell L. Stevens, "The Normalisation of Homeschooling in the USA," *Evaluation and Research in Education* 17, nos. 2 and 3 (2003): 90–100.

14. John Holt, *How Children Learn* (New York: Delacorte Press, 1982).

15. John Holt, *How Children Learn*, rev. ed. (New York: Perseus, 1995), xii.

16. Teri J. Brown and Elissa M. Wahl, *Christian Unschooling: Growing Your Children in the Freedom of Christ* (Vancouver, Wash.: Vancouver Press, 2001), 1.

17. Moore Foundation and Academy, "The Moore Formula," http://www.moorefoundation.com/article.php?id=3.

18. Home School Legal Defense Association, "About HSLDA," http://www.hslda.org/about/default.asp.

19. Scott W. Somerville, "The Politics of Survival: Home Schoolers and the Law," Home School Legal Defense Association, http://www.hslda.org/docs/nche/000010/PoliticsOfSurvival.asp.

20. Paula Rothermel, "Can We Classify Motives for Home Education?" *Evaluation and Research in Education* 17, nos. 2 and 3 (2003): 74–89.

21. Stevens, "Normalisation of Homeschooling in the USA."

22. Bauman, "Home Schooling in the United States."

23. Stevens, "Normalisation of Homeschooling in the USA."

24. See BBC News, "Home Learning Numbers Uncertain," February 23, 2007, http://news.bbc.co.uk/2/hi/uk_news/education/6389211.stm; and Matthew Charles, BBC News, "Growth Market in Home Education," March 18, 2005, http://news.bbc.co.uk/2/hi/uk_news/education/4362145.stm.

25. Amanda Petrie, "Home Education in Europe and the Implementation to Changes in the Law," *International Review of Education* 47, no. 5 (2000): 477–500.

26. U.S. Department of Education, National Center for Education Statistics, "Parent and Family Involvement in Education: Survey of the 2003 National Household Education Surveys Program," http://nces.ed.gov/programs/quarterly/vol_7/1_2/4_9.asp; Donald MacLeod, "Bullying Main Reason for Home Educating, Says Survey," *Guardian Unlimited*, August 8, 2005, http://education.guardian.co.uk/schools/story/0,1544914,00.html.

27. Bureau of Justice Statistics and the U.S. Department of Education, "Serious Violent Crime at School Continues to Fall," press release, http://www.ojp.usdoj.gov/bjs/pub/press/iscs06pr.htm. Full report, *Indicators of School Crime and Safety, 2006*, available at http://nces.ed.gov/pubs2007/2007003a.pdf.

28. "An educational aim must be founded upon the intrinsic activities and needs (including original instincts and acquired habits) of the given individual to be educated." John Dewey, *Democracy and Education* (New York: Macmillan, 1916).

29. Steven F. Duvall, Joseph C. Delquadri, and D. Lawrence Ward, "A Preliminary Investigation of the Effectiveness of Homeschool Instruction Environments for Students with Attention-Deficit/Hyperactivity Disorder," *School Psychology Review* 33, no. 1 (2004): 140–58.

CHAPTER THREE

1. John Taylor Gatto, "American Education History Tour," http://www.johntaylorgatto.com/historytour/history1.htm.

2. Section 7 of The Education Act 1996 (England and Wales), http://www.opsi.gov.uk/ACTS/acts1996/96056—a.htm.

3. Edith Hamilton, *The Ever-Present Past* (New York: Norton, 1964), 26–27.

4. Patrick Basham, "Homeschooling: From the Extreme to the Main Stream," Fraser Institute, (2001) http://www.fraserinstitute.ca/admin/books/files/homeschool.pdf.

5. Yasmin Tara Rammohan, "A Growing Movement: Black Homeschooling," *Austin Weekly News,* April 18, 2007, http://austinweeklynews.1upsoftware.com/main.asp?SectionID=1&SubSectionID=1&ArticleID=1183&TM=8263.

6. "Learning Slows Physical Progress of Alzheimer's Disease," *Science Daily,* January 24, 2007, http://www.sciencedaily.com/releases/2007/01/070123182024.htm.

7. Lev Grossman, "The Real-Life Boy Wizard," *Time,* August 29, 2005, 69.

8. See, for example, Andrea Neal, "Homeschoolers: Into the Mainstream," *Saturday Evening Post,* September/October 2006; and Michael H. Romanowski, "Revisiting the Common Myths about Homeschooling," *Clearing House,* January/February 2006.

9. Elizabeth Maker, "Home Schoolers Find Strength in Numbers," *New York Times,* February 11, 2007.

10. Elissa Gootman, "Taking Middle Schoolers Out of the Middle," *New York Times,* January 22, 2007.

11. Linda Silverman, "Social Development or Socialization?" http://www.gifteddevelopment.com/Articles/Social%20Development.html.

12. Anthony Giddens, Mitchell Duneier, and Richard P. Appelbaum, *Introduction to Sociology,* 5th ed. (New York: Norton, 2005).

13. John-John Williams IV, "It's Gym Time at Dance School," *Baltimore Sun,* November 11, 2006.

14. Roy Underhill, "'The Constancy of Change': Remarks at the Opening Session of WPA's 19th Annual Home Education Conference on May 4, 2002," *WPA Newsletter* 72 (2002): 7–8.

CHAPTER FOUR

1. Bruce Bower, "Grade-Schoolers Grow into Sleep Loss," *Science News* 157, no. 21 (2000): 324.

2. Siri Carpenter, "Sleep Deprivation May Be Undermining Teen Health," *Monitor on Psychology* 32 no. 9 (2001), http://www.apa.org/monitor/oct01/sleepteen.html.

3. Wisconsin Parents Association, *Homeschooling in Wisconsin: At Home with Learning*, 6th ed. (Madison, Wis.: Wisconsin Parents Association, 2006), 189.

4. Denise Clark Pope, *"Doing School": How We Are Creating a Generation of Stressed-Out, Materialistic, and Miseducated Students* (New Haven: Yale University Press, 2003).

5. "Chat Wrap-Up: Self-Directed Learning," *Education Week* 26, no. 16 (2006): 33.

6. International Society for Arachnology, www.arachnology.org.

7. "What Is Classical Education?" The Well-Trained Mind, http://www.well-trainedmind.com/classed.html.

8. Susan Wise Bauer, *The Well-Educated Mind: A Guide to the Classical Education You Never Had* (New York: Norton, 2003).

9. As of May 6, 2007, http://groups.yahoo.com/search?query=homeschool.

10. Grace Llewellyn and Amy Silver, *Guerilla Learning: How to Give Your Child a Real Education With or Without School* (New York: John Wiley and Sons, 2001).

11. Ned Johnson and Emily Warner Eskelsen, *Conquering the SAT: How Parents Can Help Teens Overcome the Pressure and Succeed* (New York: Palgrave Macmillan, 2007), 38.

12. Amanda Petrie, "Home Education in Europe and the Implementation of Changes to the Law," in *International Review of Education,* Unesco Institute for Education (Netherlands: Kluwer Academic Publishers, 2001), 479.

CHAPTER FIVE

1. Lenore Colacion Hayes, *Homeschooling the Child with ADD (or Other Special Needs): Your Complete Guide to Successfully Homeschooling the Child with Learning Differences* (Roseville, Cal.: Prima, 2002).

2. Brock Eide and Fernette Eide, *The Mislabeled Child: How Understanding Your Child's Unique Learning Style Can Open the Door to Success* (New York: Hyperion, 2006), 5–6.

3. Jeffrey Zalow, "Those Afflicted with ADHD Are Often the Most Creative," *Wall Street Journal,* February 3, 2007.

4. Martha Kennedy Hartnett, *Choosing Home: Deciding To Homeschool with Asperger's Syndrome* (London: Jessica Kingsley, 2004).

5. "Understanding Your Child's Learning Style," Questions and Answers with Brock Eide and Fernette Eide, *The Washington Post*, August 18, 2006, http://www.washingtonpost.com/wp-dyn/content/discussion/2006/08/11/DI2006081100706.html

6. Lisa Pyles, *Hitchhiking Through Asperger Syndrome* (London: Jessica Kingsley, 2001), 57–58.

7. James T. Webb, et al., *A Parent's Guide to Gifted Children* (Scottsdale, Ariz.: Great Potential Press, 2006).

8. Ruthe Lundy, *Dimensions of Learning for the Highly Gifted Student* (Palo Alto, Cal.: Palo Alto Unified School District), ERIC Document Reproduction Service No. ED 155 864, 1978.

9. Alvero Sanchez, et al., "Patterns and Correlates of Physical Activity and Nutrition Behaviors in Adolescents," *American Journal of Preventive Medicine* 32, no. 2 (February 2007).

10. Elizabeth and Dan Hamilton, *Should I Home School?* (Downers Grove, Ill.: InterVarsity Press, 1997).

CHAPTER SIX

1. Cafi Cohen, *Homeschoolers' College Admissions Handbook* (Roseville, Cal.: Prima Publishing, 2002).

2. Michael Winerip, "Young, Gifted, and Not Getting into Harvard," *New York Times*, May 6, 2007.

3. Gene Maeroff, *A Classroom of One: How Online Learning Is Changing Our Schools and Colleges* (New York: Palgrave Macmillan, 2003).

4. David Albert, *And the Skylark Sings with Me: Adventures in Homeschooling and Community-Based Education* (British Columbia, Canada: New Society Publishers, 1999).

5. Cafi Cohen, *Homeschooling: The Teen Years* (Roseville, Cal.: Prima Publishing, 2000).

6. Alison McKee, "Fly-Fishing to College," *Home Education Magazine*, March-April 1998, http://www.homeedmag.com/HEM/HEM152.98/152.98_art_fly.clg.html.

7. North Star: Self-Directed Learning for Teens, http://www.northstarteens.org/meta.cfm?gpt=4&mt=13.

CHAPTER SEVEN

1. World Book Encyclopedia and Learning Resources, http://www.worldbook.com/wb/Students?curriculum

2. Sue Shellenbarger, "Extreme Juggling: Parents Home-School the Kids While Holding Full-Time Jobs," *Wall Street Journal*, September 14, 2006, sec. D.

CHAPTER EIGHT

1. U.S. Department of Education, National Center for Education Statistics, "Parent Survey of National Household Education Surveys Program," 1999.

2. Jeffrey Freed and Laurie Parsons, *Right-Brained Children in a Left-Brained World: Unlocking the Potential of Your ADD Child* (New York: Simon and Schuster, 1997).

CHAPTER NINE

1. Wisconsin Parents Association, *Homeschooling in Wisconsin: At Home with Learning,* 6th ed. (Madison, Wis.: Wisconsin Parents Association, 2006).

2. Ibid.

3. Daniel Golden, "Mandatory Testing? Not Quite," *Wall Street Journal,* classroom edition, March 2003, http://wsjclassroomedition.com/archive/03mar/EDUC.htm.

4. Wisconsin Parents Association, *Homeschooling in Wisconsin.*

INDEX